T0205579

Springer Aerospace Technology

Series Editors

Sergio De Rosa, DII, University of Naples Federico II, NAPOLI, Italy

Yao Zheng, School of Aeronautics and Astronautics, Zhejiang University, Hangzhou, Zhejiang, China

Elena Popova, AirNavigation Bridge Russia, Russia, Russia

The series explores the technology and the science related to the aircraft and spacecraft including concept, design, assembly, control and maintenance. The topics cover aircraft, missiles, space vehicles, aircraft engines and propulsion units. The volumes of the series present the fundamentals, the applications and the advances in all the fields related to aerospace engineering, including:

- structural analysis,
- aerodynamics,
- aeroelasticity,
- aeroacoustics,
- flight mechanics and dynamics
- orbital maneuvers,
- avionics,
- systems design,
- materials technology,
- launch technology,
- payload and satellite technology,
- space industry, medicine and biology.

The series' scope includes monographs, professional books, advanced textbooks, as well as selected contributions from specialized conferences and workshops.

The volumes of the series are single-blind peer-reviewed.

To submit a proposal or request further information, please contact:

Mr. Pierpaolo Riva at pierpaolo.riva@springer.com (Europe and Americas)
Mr. Mengchu Huang at mengchu.huang@springer.com (China)

The series is indexed in Scopus and Compendex

More information about this series at https://link.springer.com/bookseries/8613

Ji Wu

Introduction to Space Science

Ji Wu
National Space Science Center
Beijing, China

Translated by
Yongjian Xu
National Space Science Center
Beijing, China

Qingjiang Bai
National Space Science Center
Beijing, China

ISSN 1869-1730   ISSN 1869-1749 (electronic)
Springer Aerospace Technology
ISBN 978-981-16-5753-5   ISBN 978-981-16-5751-1 (eBook)
https://doi.org/10.1007/978-981-16-5751-1

Jointly published with Science Press
The print edition is not for sale in China Mainland. Customers from China Mainland please order the print book from Science Press.

Translation from the Chinese language edition: *Introduction to Space Science* by Ji Wu, © China Science Publishing & Media Ltd (Science Press) 2020. Published by China Science Publishing & Media Ltd (Science Press). All Rights Reserved.

This Springer imprint is published by the registered company Springer Nature Singapore Pte Ltd.
The registered company address is: 152 Beach Road, #21-01/04 Gateway East, Singapore 189721, Singapore

Foreword

Space science adopts the spacecraft as the main tools to carry out scientific research and experiments, which is an integral part of the space endeavor. For China, with the gradual increase of comprehensive national power and the emergence of science and technology innovation as the main driving force of development, it is the inevitable choice, to vigorously promote the development of space science, which will trigger a wave of intensive demands for space technology, and provide theoretical guidance and support for the development of space applications.

I have known Professor Ji Wu, the author of this book, for a long time. As the former Director General of the National Space Science Center (NSSC) of the Chinese Academy of Sciences (CAS) and the President of the Chinese Society of Space Research (CSSR), one of the key players in promoting the development of space science in China in recent years, he has done substantial work and made tremendous contributions to the development of space science in China. This book is based on the development of China's space programs and Prof. Ji Wu's personal experience in so many years, and at the same time draws on the practices of spacefaring nations and institutions, which makes a good reading for scientists, engineers, and project managers. This book can also be used as a textbook and reference for related courses in universities and colleges.

For young students who are new to the space sector, they need to get familiar with the origin and background of the space programs. This book traces back to the early days of human observation of space and cosmos, reviews the development of science to reveal the asperation intention of space exploration, and introduces the history of space technology with the focus on launch vehicles which helps to overcome the gravity of the Earth to enter space. This will greatly broaden the horizons of these young students.

For the people with engineering background in space sector, Chaps. 3 and 4 are very unique in that after a comprehensive and macroscopic introduction to several important branches of space science, the key scientific frontiers of each branch are briefly described, which can greatly stimulate their curiosity. Besides, for people who are no longer engaged in scientific research, and only participate in these space science missions as engineers or managers, they will be proud of doing their bit in the exploration of such interesting scientific frontiers. Therefore, the introduction of this background knowledge is essential for all those involved in space science missions.

After presenting the necessary background knowledge of space exploration, this book provides a systematic overview of the major systems of space technology, highlighting the orbit, launcher, and spacecraft/satellite systems, with particular focus on the relationship between payloads and orbits, launcher and spacecraft in scientific missions. This knowledge is essential for both engineers and scientists who are involved in space science missions. For engineers, they need to understand the requirements of a space science mission and the payloads, while for scientists, they need to know the engineering constraints and boundaries of orbits, launch vehicles, and spacecraft.

This book differentiates itself from purely scientific and technical monographs with substantial discussions on space science mission management. In our long time cooperation, Prof. Ji Wu has been responsible for specific technical work, as well as important management work in many major space missions, especially space exploration missions. Therefore, he has rich experience in systems engineering. The experience in management issues introduced in this book is based on the 60 years' development of Chinese space program, as well as his personal experiences in leading the implementation of major space science missions. As the reading unfolds, the reader can have the panoramic view, from a mission manager's perspective, of the entire process of a science mission from mission planning, pre-research, selection, approval, development, launch, operations, and evaluation, with the prevailing principle of maximizing scientific output, which is articulated by the author to the fullest extent. I believe that those engaged in space science missions would benefit greatly from reading this book, which would facilitate their future research work.

Finally, the author discusses the relationship between space science, space technology, and space applications in very concise language and with the help of diagrams. It can be seen that these three integral aspects are mutually dependent, mutually supportive, and mutually reinforcing.

All in all, this book makes a fascinating reading. It is a summary and overview of the development of the space endeavor in China in the last 60 years. I hope that the present and future scientists, engineers, and project managers of space science missions could read the book, which can also serve as an introductory textbook for young students who will enter the space sector. It is hoped that the publication of this book will inject new momentum into the space science endeavor in China and promote its faster, better, and sustainable development.

Beijing, China   Academician Peijian Ye
May 2020

Preface

Although humans have been observing and recording the stars and the cosmos for thousands of years, it is only in the last 60 years or so that we have, in the real sense, entered the space age. Space science is an emerging interdisciplinary field, which thrives from the development of space technology and uses the spacecraft as the tools to conduct research in space. Compared with other traditional disciplines, space science, as a basic research area, is government dominated and its research directions are planned accordingly. Since its expected scientific output may lead to discoveries and breakthroughs in major basic science frontiers, it belongs to the basic research. However, the implementation of a space science mission requires integration of various disciplines and systems engineering technology, especially the space technology, and relies on the spacecraft as the platform of observation and experiment to achieve scientific objectives. In addition, due to the uniqueness of scientific discoveries, the requirements on the space technologies by a space science mission are constantly becoming higher and higher, leading to innovations and upgrading in space technology. Transforming these new technologies into applicable technologies on the ground can even give rise to new strategic industries and drive the development of economy and society. In this respect, it is also a part of national space activities with the potential to enhance the space technology capacity.

Traditional basic science education based on individual disciplines aims to train professionals in specific disciplines, such as physicists, astronomers, space physicists, planetary scientists, atmospheric and ionospheric physicists, solid Earth physicists, and life scientists. In the traditional curriculum, seldom do they have the opportunity to learn related knowledge of large systems engineering, such as the space systems engineering. But in the future, when the abovementioned professionals are engaged in space science research, they will turn into space scientists, who need to understand orbital dynamics, space environment, various space vehicles like spacecraft/satellites, as well as knowledge of systems engineering management. In the traditional education system to train the engineers, be it mechanical, electrical, or material, the focus of education is on feasibility, reliability, repeatability, and implementability, and the graduates can directly participate in the engineering implementation. But when involved in a space science mission, these engineers often do not understand the language of the scientists who are striving to explore the unknown, and they are unwilling to accept the engineering and technical challenges and risks brought about by the new requirements. However, prioritizing the scientific objectives prevails in a space science mission, from mission proposal to operations, and science and engineering are closely bonded together. On one hand, scientists without basic knowledge of space technology and systems engineering cannot communicate with the mission development and research team, and are incapable of coordinating the mission development with hundreds of participants, which leads to the failure of realizing their dreams. On the other hand, engineers who fail to understand the language of science cannot understand the requirements of scientific detection and observation, and cannot make reasonable improvements to the engineering design to best meet the needs of the scientific objectives.

This book, from a top-down, interdisciplinary, and systematic perspective, aims to provide a systematic introduction of the knowledge on the frontiers of various branches of space science disciplines, space technology, and systems engineering, to highlight the characteristics of space science missions as compared with other space missions, to lay the fundamental systematic knowledge for scientists and engineers who wish to engage and participate in space science missions in the future, and to cultivate scientists and engineers as potential principle investigators, chief designers, and project managers.

The outline of this book is as follows: Chaps. 1 and 2 focus on the reasons to conduct research in space and the history of space exploration; Chaps. 3 and 4 introduce the major frontier issues in space astronomy, planetary science, space solar physics, space physics, space Earth science, microgravity science, and space life sciences, respectively; Chap. 5 introduces the space systems engineering and its systems; Chaps. 6–8 introduce the technical foundations for space science missions, including orbit, attitude and TT&C, scientific payloads and its application environment, mission planning, and operations. Chaps. 9–12 focus on the key factors of space science mission management, including mission proposal and its selection, mission development and the duty of scientists and engineers, quality management and risk control, full mission lifecycle management, and output evaluation. Chap. 13 introduces the international cooperation in space science missions; Chap. 14 introduces the space science programs in China; Chap. 15, as the wrap up of the book, not only gives definitions to space science, space technology, and space applications, but also discusses their relationships.

This book is based on my two-semester course for the graduate course of the School of Astronomy and Space Science, University of the Chinese Academy of Sciences. This book received substantial support from Associate Prof. Bai Qingjiang, the course assistant, as well as the necessary help from Prof. Zheng Jianhua from the National Space Science Center of the Chinese Academy of Sciences and Alvaro Gimenez, former Science Director of the European Space Agency. My thanks also go to colleagues from several departments of the National Space Science Center for their assistance, including the Space Science and Deep Space Exploration Study Center, the Space Science Program Center, and the Space Science Mission Operation and Control Center.

My thanks go to the translators of this book for their devoted effort to bring it into the current shape. Mr. Xu Yongjian is responsible for the translation of Chaps. 1–8 and Chap. 13, and Ms. Bai Qingjiang is responsible for Chaps. 9–12 and Chaps. 14 and 15.

Finally, my gratitude also goes to Zhu Pingping, the Editor of the Science Press, for her hard work, which has enabled efficient publication of this book.

Beijing, China   Ji Wu
April 2020

Contents

About the Author

Prof. Ji Wu is President of the Chinese Society of Space Research (CSSR) and former Director General of the National Space Science Center (NSSC) of Chinese Academy of Sciences (CAS). He is full member of IAA (International Astronautics Academy), fellow of the IEEE Geoscience and Remote Sensing Society, member of the Advisory Board of Luxembourg Government for Space Resources, and member of the Advisory Committee of UAE Space Agency. He once served as Vice-President of the Committee for Space Research (COSPAR) (2010–2018), Head of the Strategic Priority Program on Space Science of CAS, Chief Designer of the application system of Double Star Program, Principle Investigator of Yinghuo-1, and Project Manager of Scientific Payload System of Chinese Lunar Exploration Program Chang'e-1 and Chang'e-3.

Chapter 1
Reasons to Conduct Research in Space

1.1 Introduction

What are the reasons to conduct research in space? For many disciplines, even including astronomy, research could be carried out on the ground. For example, Galileo Galilei (1564–1642) pioneered the practical ground observation of celestial bodies using telescopes. Another example is the employment of ground-based radars to observe and study the ionosphere. Even so, there's still a lot of research that can't be done on the ground, which necessitates the research in space. This chapter will focus on the reasons to go into space.

From the beginning of the space age, the fundamental and foremost objective of entering space to carry out research is to unveil the mystery of space and increase our knowledge of space. Before the launch of the first artificial satellite in 1957, the outer space reaching beyond the atmosphere is shrouded in mystery, where the neutral atmosphere thins out and is ionized by the ultraviolet light from the Sun when reaching further out, hence creating the ionosphere. But, questions remain to be answered, e.g., how the electrons and ions in the ionosphere are distributed and how do they move? What effect does the Earth's magnetic field exert on these charged particles?

After gaining access to space, for the first time in human history, we have the opportunity to observe the Planet Earth from hundreds or even thousands of kilometers away. When we observe it from such a distance, our perceptions of the Earth become very different. The changes that the Earth presents to us become systematic, such as the formation and movement of typhoons.

What's more, once break free the obstacles of the atmosphere, we have the liberty to make full use of the resources of the entire electromagnetic spectrum. Previously, the low-frequency electromagnetic waves, terahertz, and infrared wavelengths, as well as wavelengths beyond the ultraviolet that are normally blocked by the atmosphere. Entering into space enables us to observe the universe in full electromagnetic spectrum.

© Science Press 2021
J. Wu, *Introduction to Space Science*, Springer Aerospace Technology,
https://doi.org/10.1007/978-981-16-5751-1_1

For an in-orbit spacecraft, the centrifugal force generated by its rotation around the Earth is offset by the gravitational force of the Earth, providing us an equivalent microgravity environment for a long period of time. Previously, our understanding of the kinetic properties of matter and the rule of life activity is actually based on the condition of the gravity of the Earth. So, if we remove the gravity, will the movement of matter and life remain the same?

In short, gaining access to space is to enter a larger laboratory where the experiments previously impossible on the ground can be carried out.

1.2   To Explore the Unknown Space Environment

Before the space age, the human knowledge of space was limited to speculations and theoretical conjectures. The atmosphere thinned out, but then what? The ultraviolet light from the Sun ionizes atoms in the atmosphere, allowing electrons to escape and correspondingly form the ionosphere. The answer was not clear in 1901, when Guglielmo Marconi (1874–1937), an Italian radio engineer, successfully transmitted a radio signal across the Atlantic. Marconi was puzzled for a long time by the fluctuation of radio waves, which apparently traveled a winding path to reach the destination thousands of kilometers away.

We now know that, for a transmission distance of more than 5000 km from the west coast of Europe to the east coast of the United States, the radio waves reached the receivers with the help of ionospheric reflections. It turns out that Marconi's first successful transoceanic radio communication in 1901 verified the existence of the ionosphere.

The human knowledge about ionosphere stops there. By that time, we still didn't know where the upper boundary of the ionosphere is, or how positively charged ions and negatively charged electrons in the ionosphere behave. Only after 1957 did the answers to these questions become clear. Therefore, to study the unknown space environment is the core of space research. This is especially the case for the first artificial satellite launched by the Soviet Union on October 4, 1957, and the first American artificial satellite launched on January 31, 1958. Malfunctions were detected on the instruments for both satellites and American scientists tended to believe that the malfunction is not due to the instrument itself but rather to the existence of intense high-energy particle zone in the near Earth space, which was later identified and consequently named as Van Allen belt. This is the first major discovery in the space history of mankind.

1.3   To Break Free the Barrier of Atmosphere
to Electromagnetic Wave

Since Galileo pointed his telescope into space, human beings have broken the limitations of space observation with the naked eye and began to use scientific instruments to observe the universe. The spectrum of electromagnetic waves we can observe

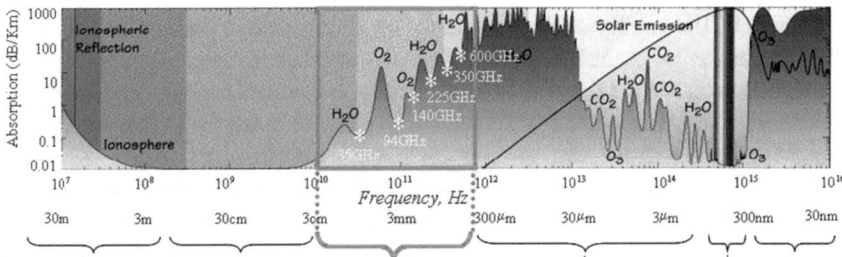

Fig. 1.1 Atmospheric absorption across the electromagnetic spectrum

was then gradually expanded to radio waves. However, the protective atmosphere proves to be an obstacle for the ground-based observations using ultraviolet and low-frequency radio spectrum. Figure 1.1 shows the atmospheric absorption spectrum diagram, in which the ordinate is the atmospheric absorption expressed in decibels loss per kilometer.

Since 1957, humans began to break the atmospheric barrier by placing observational instruments on satellites. New images of the universe and the Sun were obtained in wave lengths such as infrared, ultraviolet, X-ray, and low-frequency electromagnetic waves. Since then, space astronomy and space solar physics flourished as individual disciplines in their own names.

In addition, the atmosphere can absorb the electromagnetic waves, which, in turn, makes it possible to carry out space-based observation of the physical characteristics of the Earth's atmosphere. For example, the frequency in the vicinity of the temperature absorption line can be used to observe the distribution of atmospheric temperature at different altitudes, and the frequency in the vicinity of the water vapor absorption line can be used to observe the distribution of atmospheric water vapor, etc. These technological breakthroughs have promoted the development of space Earth science.

1.4 To Utilize the Orbital Altitude Resources

From the mountain top, you can see farther. In general, the altitude of satellite orbit is above 500 km. Geosynchronous Earth Orbit (GEO) is as high as 36,000 km. The natural field of view of the human eye is about 45°, which is also the field of view of a standard camera. Accordingly, for Earth observation, the width of more than 400 km can be obtained at an altitude of 500 km (the width covered by push-sweep camera on an operating satellite is called swath), and cities such as Beijing, Moscow, and New York can be seen in a panoramic view. If the Earth is observed in Geosynchronous Earth Orbit, the field of view of the entire Earth is less than 18°. By designing remote sensors with different field of views, we can obtain ground images with different swaths, making possible the systematical observation of the Earth.

Therefore, the satellite orbit promises the unprecedented altitudes where the Earth observation can be conducted and the data obtained can be used to study the large-scale phenomena of the Earth system, such as typhoon, ocean currents, El Nino, the atmospheric pollution caused by volcanic eruptions, and even the global water cycle, biosphere, energy cycle, ice and snow cycle and lithosphere, etc. This provides the most important observation platform for space Earth science to study the Earth as a system. After the technical realization of putting a satellite in Geosynchronous Earth Orbit, human beings can continuously and comprehensively monitor the changes of the Earth.

1.5   To Unveil the Mystery of the Earth's Gravitational Field

We live on the Earth under the effect of 1G gravitational field. To put it in a figurative way, all the kinetic properties of matter and the rule of life activity lie beneath the veil of 1G gravitational field on the surface of the Earth.

Using space as platform, it is possible to carry out on-board scientific experiments that cannot be done otherwise on the ground. Among these on-board experiments, the microgravity science experiments [1] are the most prominent ones. The centrifugal force generated by the spacecraft as it orbits the Earth offsets the Earth's gravity, hence creating a continuous and stable artificial microgravity environment. Such an environment may reveal the kinetic properties of matter, which are impossible to be discovered due to the effect of gravity. The studies of the laws of physics, e.g., the law of fluid physics, combustion, and semiconductor material growth, are collectively known as microgravity science.

In addition to the study of laws of physics, some fundamental issues in life science can also be examined in space microgravity environment, such as the cultivation of cells and plants, which gives birth to space life science. Of course, when studying the life science issues, consideration should be given to the effects of space particle radiation and the influence of weakened Earth's magnetic field.

With the increase of manned space activities, the required period of stay for astronauts in space becomes longer and longer. Can animals and human beings live in space for a long time? These new scientific questions posed by man's entry into space can also be deemed as part of space medicine.

1.6   To Make Full Use of Other Aspects of Space Environments

There are other resources in space that are not available on the ground, such as radiation.

Without the protective atmosphere, the intensity of cosmic rays is considerably higher in space than that on the Earth's surface. With the effect of Earth's magnetic field in full play, the high-energy particles of the solar wind will concentrate in certain regions of the terrestrial space. Combined with the particles of the Earth's radiation belt, a unique particle radiation environment is formed featuring a wide range of energy spectrum, high flux, and continuity, which cannot be simulated on the ground. This is of unique significance for life science research, such as space breeding, etc.

In addition, since the Earth's diameter is only more than 12,000 km, the longest interferometry baseline we can obtain between two stations on the ground is no more than the Earth's diameter. To get a longer interferometry baseline for research like gravitational wave detection or interference imaging observation in radio astronomy, we must venture into space to place spacecraft there in a bid to form an interferometry baseline of millions of kilometers.

Compared with the surface of the Earth, a better electromagnetic radiation environment can be obtained in space. The far side of the moon, for example, shields various man-made and natural electromagnetic radiation from the Earth, making it the ideal place for low-frequency electromagnetic observation of the universe.

In addition, it is easier to obtain extremely low-temperature and high-vacuum environment in space, especially in orbits beyond the Earth orbits.

1.7   Definition of Space Science

With spacecraft as the main tools, space science is defined as the study of natural phenomena and their underlying rules in physics, astronomy, chemistry, and life science which exist in solar-terrestrial space, interplanetary space, and even the universe as a whole. The following shows the definitions of space science in different stages of its historical development, as well as its ever-expanding disciplines.

Prior to the space age, space science was defined in a narrow sense as the study of space, such as the upper atmosphere, the ionosphere, and the distribution of the Earth's magnetic field in space.

In the early days of space age,[1] space science is defined as the study of the space physics phenomena surrounding the spacecraft, e.g., charged particles, neutral particles, electromagnetic field distribution, and its patterns of changes. The research in early days has improved our understanding of the electromagnetic field and particle distribution in the magnetosphere, which is controlled by the Earth's magnetic field. New concepts are proposed, such as the magnetopause, bow-shock, magnetic tail, polar cusp, South Atlantic Anomaly, etc. The period also saw the march from terrestrial space to the solar system, the birth of planetary science, and the emergence of comparative planetary science through the comparative study of the composition of the Earth and other planets.

Five years into the space age, scientists began to use spacecraft as platforms to conduct astronomical observations, and *space astronomy* and *space solar physics* were established as disciplines accordingly.

Ten years into the space age, scientists began to study the Earth with remote sensors of various frequency bands, which initiated the discipline of *space Earth science*.

From the late 1960s and mid-1970s, humans landed on the moon, which is followed by the launch of space laboratories, space shuttles, and manned space stations. A large number of scientific experiments held in microgravity environments are known as *microgravity science* and *space life sciences* research [2]. The research that aims at the physiological changes of people in space is called *space medicine*.

In the past 20 years, making full use of the ultra-low temperature and high microgravity level brought about by spacecraft, we began to verify the basic physical laws, and thus started the *space fundamental physics* experiment.

In addition, based on space technology, space science has opened up the field of space technology science by carrying out technology research and technology demonstration in space, such as propulsion technology, attitude control technology, navigation technology, and thermal control technology.

Table 1.1 lists the main research contents of various disciplines of space science.

---

[1] *Note:* Space age starts from October 4, 1957, when the Soviet Union successfully launched Sputnik I, the world's first artificial satellite, which took about 98 min to orbit the Earth on its elliptical path. It ushered in new era of political, military, technological, and scientific developments.

Table 1.1 Main research contents of various disciplines of space science

Discipline	Main research contents
Space astronomy	Space astronomy is a discipline that uses spacecraft to observe celestial bodies in space and studies the morphology, structure, composition, physical properties of motion, and evolution of celestial bodies. In terms of wavelengths, there are several original branches of astronomy, such as radio astronomy, infrared astronomy, ultraviolet astronomy, X-ray astronomy, gamma ray astronomy, etc. If the observations are carried out in space by spacecraft, these branches can be called space radio astronomy, space infrared astronomy, space ultraviolet astronomy, space X-ray astronomy, space gamma ray astronomy, etc
Space solar physics	Space solar physics uses spacecraft to carry out observations and studies of the Sun. According to the objects of study, the discipline incorporates the solar magnetic field, solar flares, coronal mass ejections, solar atmospheric structure and dynamics, solar medium- and long-term changes, etc
Space physics and space environment	The discipline uses spacecraft as well as ground-based observation facility to study physical phenomena in solar-terrestrial space, such as the upper atmosphere of the Sun, the movement of interplanetary space plasma, and the Earth and other planets' magnetosphere, ionosphere, and atmosphere, as well as their interactions and cause-and-effect relations; it also studies the space environment, such as the space electromagnetic environment, space charged particle environment, middle and upper atmosphere, meteoroid environment, and space debris environment; combined with applications, space physics, and space environment gave rise to new disciplines, namely, space weather and space climatology
Space Earth science	The discipline uses spacecraft to conduct systematic research of the Earth as a whole, including the movements of the atmosphere, hydrosphere (water cycle), energy cycle, lithosphere, and biosphere and their interactions. Among them, global change is the core scientific question of space Earth science research
Planetary science	The planetary science uses spacecraft to investigate planets (including comparative studies with the Earth) and their moons, planetary systems (especially the solar system), and their formation processes. The study of exoplanets' formation and their habitability has recently been integrated into the field of planetary science

(continued)

Table 1.1 (continued)

Discipline	Main research contents
Microgravity science	The discipline mainly studies the laws of equilibrium of matter and kinetic properties of matter in a microgravity environment. The main research fields include microgravity fluid physics, microgravity combustion, space materials science, and fundamental physics experiments in microgravity environment, which are the main research fields of manned spaceflight
Space life sciences	The discipline studies the phenomena, processes, and the rule of life activities under the influence of special environmental factors in space (such as microgravity, cosmic radiation, magnetic variation, vacuum, high temperature and low temperature, etc.); explores the performance and ability to survive in outer space for extraterrestrial life and human beings; and studies the origin, evolution, and rule of life
Space fundamental physics experiment	The research mainly include cold atom physics, low-temperature condensed matter physics, relativity and gravity physics, space experiments of quantum entanglement

References

1. Wenrui H (2010) Introduction to microgravity science. Science Press, Beijing
2. National Natural Science Foundation of China, Chinese Academy of Sciences (2019) Space science. Science Press, Beijing

Chapter 2
History of Human Space Exploration

2.1 Introduction

The history of space exploration begins with space observation from the ground, which is almost as long as the recorded history of mankind. In ancient times, human observation of space is out of awe, fear, or divination. The observation is also out of the necessity for farming, e.g., the observation of time, seasons, and climate. Yet these simple thoughts and ideas are more or less like groping in the darkness before Nicolaus Copernicus (or Mikolaj Kopernik in Polish, 1473–1543) proposed the heliocentric theory.

Galileo directed his telescope into space,expanding the space beyond the naked eye, which is the first time in history and introduces a new way of space observation. Since then, the frequency band visible to the naked eye has gradually expanded from visible light to the full electromagnetic spectrum.

It takes another 350 years to make the giant leap from ground observation of the universe and planets via telescopes to the launch of spacecraft into space. During this period, countless science and technology pioneers have made their contributions, among whom Qian Xuesen (also written traditionally as Tsien Hsue Shen) was an indispensable player in human's journey into space. He and Zhao Jiuzhang (also written traditionally as Jaw Jeoujang) have made pioneering contributions to the development of Chinese space programs and space science endeavors.

This chapter will touch upon the various frontiers of space technology and their general development trends.

© Science Press 2021

J. Wu, *Introduction to Space Science*, Springer Aerospace Technology,
https://doi.org/10.1007/978-981-16-5751-1_2

2.2 History of Space Exploration

What did our ancestors see by looking up into the sky with their naked eyes? The first batch of celestial bodies that greet their eyes are the Sun, the moon, and stars,[1] and their movement in the sky followed fixed periodical sequence. Therefore, the first thing that comes to people's mind is to ponder over and summarize the rules of their positions and movements.

This is how the concept of time comes into being, e.g., day, month, year, hour, minute, and second. All the ancient cultures, be it Egypt, Babylon, Greece, India, and China, developed their own independent systems about sky phenomena and time. For example, the Greek calendar is based on the solar cycle and the Chinese calendar is based on the lunar cycle.[2] In terms of timekeeping, the Chinese clepsydra/drip-vessel and the sundial, as well as the clock in the Tower of Winds in ancient Athens, Greece, are all the embodiments of human wisdom.

In addition to the summary on the periodic movements of the Sun, the moon, and stars, the most important part of the ancient study (or simply the thinking) on space is the observation and description of sky phenomena, including solar and lunar eclipses, comets, meteors, and auroras. At the ancient time, the physical principles of these phenomena are still hidden knowledge. China has the earliest and most complete records of observations of these sky phenomena. For example, China has the earliest records of sunspots and supernovae, and the record of "New Great Star" (*Xin Da Xing*) inscribed on the oracle bone actually describes an exploding supernova (shown in Fig. 2.1). The continuous observation of Comet Halley is another example. China recorded Comet Halley for 28 times from BC until 1910, but not knowing it is the same comet. Based on the systematic knowledge of motions of celestial bodies established by Johannes Kepler (1571–1630) and Isaac Newton (1643–1727), Englishman Edmond Halley (1656–1742) successfully predicted that the comet's return period would be 76 years. Later, the comet was named "Halley's Comet". Auroras normally occur at the poles region of the Earth. Although China is in the middle and low latitudes, auroras have also been recorded in China during solar eruptions throughout the history. In addition, China observed the outburst of the Crab Nebula in the Song Dynasty. Because of its very stable X-ray emission, the Crab Nebula is now used as a beacon galaxy by X-ray astronomical observers.

In ancient China, the main purpose of detailed astronomical observations and sky phenomena observations [1] is to strengthen the royal power, hence bearing little scientific significance. In order to better serve the royal power, for more than 2000 years, extensive records of observations have been made by the emperor's astronomers or star gazers (a demanding profession that requires sleeping during the day, and observing and recording at night). Some of these records are accurate, while some are hypothetical or even false. For the researchers devoted to the history of ancient Chinese astronomy, they often spend a great deal of time scrutinizing these records by extrapolating the cycles of celestial body movements to determine

---

[1] *Note:* Here the star is a vague term generally describing the visible celestial bodies.

[2] The lunar calendar is still in use today in China.

Fig. 2.1 Record of the "New Great Star" inscribed on the oracle bone unearthed at the Yin relics in Anyang, Henan Province (1300 BC)

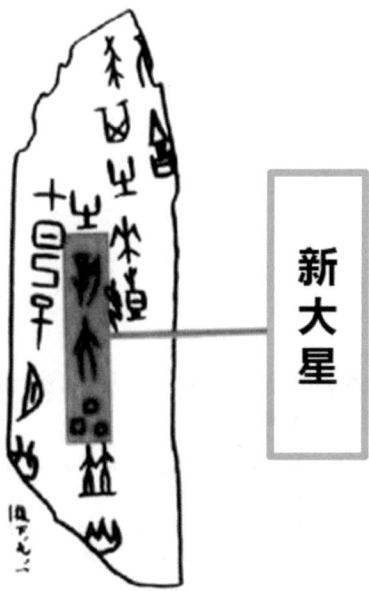

the dynasty and year. Some records are quite informative and interesting though, such as "Mars retrograde", which is interpreted in ancient China as an inauspicious sky phenomenon that would bring disasters. For this reason, Mars was traditionally known as "glittering planet" or "firefly". Now, we understand that according to the rules of celestial body movements, the Earth orbits the Sun in the inner circle, and when the Earth catches up with the Mars that orbits in the outer circle, the ground observer would find the occurrence of "Mars retrograde".

Although the observation of sky phenomena in ancient China doesn't bear much scientific significance, Chinese people's philosophical thinking of celestial bodies and universe has an early origin, such as the idea of "round heaven and square Earth" [2], the "opening of the universe by Pangu" (a giant in Chinese fable stories who opened the universe), and the *"Heavenly Questions"* by Qu Yuan (a Chinese poet in the warring states period who is regarded as the greatest patriot). In addition, the deep thinking of the Chinese people and their summary of space and geography at that time are also reflected in the *Book of Changes* and the *Eight Diagrams*.

In other civilizations, space-related thinking also started more than 2000 years ago, such as the dispute between the heliocentric theory of ancient Greece (Aristarchus, 315–230 BC) and the geocentric theory (Claudius Ptolemy, 90–168). The heliocentric theory was initially untenable due to the lack of scientific evidence. The geocentric theory, supported by the European church, dominated the human perception of the world for a long period. The heliocentric theory began to be widely accepted and became dominant in human perception of the solar system only after Copernicus deduced its correctness in theory and Galileo proved its correctness with practical

Fig. 2.2 Sketch of
Copernicus

observations. The following is the introduction to several Western pioneers and their achievements.

Nicolaus Copernicus (as shown in Fig. 2.2) is a Polish astronomer who came up with the heliocentric system (Sun-centered system) and calculated in detail the orbits of the Earth and several other planets. Under the threat of the geocentric church, his *Six Books Concerning the Revolutions of the Heavenly Orbs* was not published until his death in 1543. In fact, Copernicus initially believed in the geocentric theory, but doubts arose when he was doing calculation. Heliocentric theory changed our view of the universe and offered a scientific explanation of the movements of celestial bodies in the solar system, hence laying the foundation for modern astronomy and space science. Copernicus is one of the giants of the European Renaissance, whose remains was reburied in Vronburg Cathedral, Poland, on May 22, 2010.

Galileo Galilei (as shown in Fig. 2.3) is an Italian physicist and astronomer, whose achievements include improving telescope and the astronomical observations with telescope. He is also known for his advocacy in Copernican heliocentric theory. In 1609, Galileo, for the first time, directed his telescope to space. He discovered that the surface of the moon was uneven and made the first map of the moon's surface. On January 7, 1610, Galileo discovered four moons orbiting the Jupiter, which is the solid evidence for the heliocentric theory. This also marks the initial victory for Copernicanism. With the telescope in hand, Galileo made continuous discoveries of

Fig. 2.3 Sketch of Galileo

a ring surrounding the Saturn, sunspots, the rotation of the Sun, the phases of the Venus and the Mercury, the diurnal libration (or parallactic libration) and circumlunar libration of the moon, as well as the fact that the Milky Way is made up of countless stars. These discoveries usher in a new era for astronomy. In honor of Galileo's achievements, Jupiter's moons Io, Europa, Ganymede*, and Callisto are collectively called the Galilean moons.

Johannes Kepler (as shown in Fig. 2.4) is a German astronomer, physicist, and mathematician. An ardent supporter for Copernicus's heliocentric theory, Kepler discovered, even before Newton, *three laws of planetary motion*: the orbit law, the area law, and the harmonic law. The three laws can be described as follows: (1) the planets move in elliptical orbits with the Sun at one focus (*the orbit law*); (2) the time necessary to traverse any arc of a planetary orbit is proportional to the area of the sector between the central body and that arc (*the area law*), which means the planet in question travels faster around the perigee and travels slower around apogee; and (3) there is an exact relationship between the squares of the planets' periodic times and the cubes of the radii of their orbits (*the harmonic law*). These three laws were later confirmed by Newton's law of universal gravitation.

Isaac Newton (as shown in Fig. 2.5) is an English physicist, President of the Royal Society, and the author of *Mathematical Principles of Natural Philosophy*

Fig. 2.4 Sketch of Kepler

Fig. 2.5 Sketch of Isaac Newton

(*Philosophiae Naturalis Principia Mathematica*) and *Optics*. In his 1687 treatise, he described the three laws of motion, including the law of universal gravitation. These descriptions laid the foundation for the scientific view of the world of physics for the next three centuries and became the basis for modern engineering. By demonstrating the consistency between Kepler's laws of planetary motion and his law of gravitation, he showed that the motion of terrestrial and celestial bodies followed the same laws, which provided strong theoretical support for the heliocentric view of the universe and contributed to the revolution in science.

We can see that Copernicus and Newton are almost two centuries apart, and their findings and discoveries lay the foundation for modern astronomy and modern space science.

2.3   The Technology Advancement of Ground-Based Observations Since Galileo

The inventor of telescope is Hans Lippershey (1570–1619), a Dutch spectacle merchant. But he did not use the telescope to observe space. When Galileo learned about this invention, he immediately came up with the idea of using it to observe space.

Galileo began making telescopes in 1609. The first telescope was capable of magnifying up to 32 times. Using this telescope (Fig. 2.6), he discovered craters on

Fig. 2.6 Galileo's telescope, the first telescope ever pointed to space. Collection of Galileo Museum

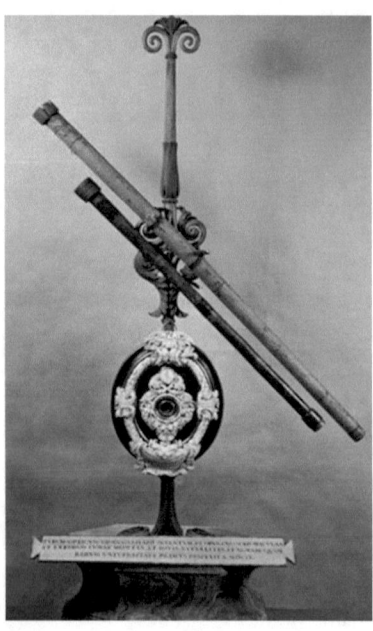

the moon, a ring of the Saturn, and four moons of the Jupiter. Thanks to telescopes, the observation capacity of human eye improved from 6 to 9 magnitudes (based on the aperture of Galileo's first telescope).

Magnitude measures the brightness of a star or other celestial body as seen by an observer on the ground. The apparent magnitude can be negative. The brighter the object, the lower the number assigned as a magnitude, and vice versa. For example, the apparent magnitude of the moon is around -12. The naked eye is capable of observing stars of 6 magnitudes. One magnitude is defined as a ratio of brightness of 2.512 times, e.g., one magnitude lower means 2.512 times bright. The absolute magnitude difference is 1 and the luminosity difference is 2.512 times. Magnitudes don't reflect the true luminosity of the star itself, because magnitudes do not take into account the distance to the star. For stars with the same level of luminosity, the more distant from an observer, the lower the apparent brightness.

Brightness is determined by the flow of photons received by the eyes, so if the aperture of the telescope's objective lens (the space-facing end) is larger than the pupils of our eyes, more photons will be received, which will be converged to the eyepiece and received by the observer. The larger the aperture, the lower the magnitude for the observer (and the more magnitudes). European Extremely Large Telescope (E-ELT), the largest ground-based astronomical telescope currently being designed, has an aperture of 39 m. The space telescope with the largest aperture was the James Webb Space Telescope (JWST), whose aperture is up to 6.5 m.

The larger aperture not only enhances the observation capacity, but also improves the spatial resolution. The aperture of Galileo's telescope is much larger than the pupils of the eyes, hence improving spatial resolution, which explains why he easily observed the craters on the surface of Moon and the ring system of Saturn. The schematic relationship between telescope aperture D (m), optical wavelength λ (m), and angular resolution θ (°) is given by Eq. (2.1).

$$\theta = 180\frac{\lambda}{\pi D} \tag{2.1}$$

It is fair to say that the invention of the Galileo astronomical telescope opened up the modern space observation, and also laid the technical foundation for the modern space studies. The observation technology, be it at any electromagnetic wave band, is based on this basic principle.

Michael Faraday (1791–1867) (as shown in Fig. 2.7) is an English physicist whose discoveries laid the foundation for electromagnetism. On October 17, 1831, Faraday discovered electromagnetic induction for the first time and developed a method to generate alternating current. On October 28, 1831, Faraday invented the first electric motor and dynamo.

Because of his great contributions to electromagnetism, he was called *the father of electricity* and *the father of alternating current*.

James Clerk Maxwell (1831–1879) (as shown in Fig. 2.8) is a British physicist and mathematician. Based on existing research by scientists like Michael Faraday, James

Fig. 2.7 Sketch of Michael
Faraday

Fig. 2.8 Sketch of James
Clerk Maxwell

Clerk Maxwell put forward classical electromagnetic field theory. He is the founder of classical electrodynamics. His *Treatise on Electricity and Magnetism* (1873) is considered the most important physics classic after Newton's *Mathematical Principles of Natural Philosophy*. The book describes the basic theory of electromagnetic waves and lays the foundation for all applications up to date that relate to electromagnetic waves.

Heinrich Rudolf Hertz (1857–1894) (as shown in Fig. 2.9) is a German physicist who proved the existence of electromagnetic waves through experiments in 1888. His research has stunned the scientific community, proving for the first time that electromagnetic waves can travel, transmit information and energy, and the traveling speed is the speed of light. The international unit of radio frequency, Hertz (Hz), is named after him. 1 Hz equals one cycle per second.

Guglielmo Marconi (as shown in Fig. 2.10) is an Italian radio engineer, entrepreneur, and founder of practical wireless telegraph communications. During his study at the University of Bologna, he successfully carried out radio communication experiments at a distance of about 2 km with electromagnetic waves. In 1901, he completed his

Fig. 2.9   Sketch of Heinrich
Rudolf Hertz

Fig. 2.10   Sketch of
Guglielmo Marconi

transatlantic wireless communication and discovered the ionosphere. In 1897, he
founded the Wireless Telegraph and Signal Company, Ltd. (changed to Marconi's
Wireless Telegraph Company, Ltd. in 1900). He won the Nobel Prize in physics in
1909 and was regarded as *the father of radio.*

Starting from Faraday, the electromagnetic wave theory became an established
theory, which provides modern scientific basis for using radio wave to observe space.
In the 1930s, two ultrashort wave antenna engineers of the Bell telephone company
first discovered electromagnetic radiation in the sky. Theoretical physicists at the time
analyzed that the very high-frequency radiation was left after the Big Bang of the
universe, and that over 13.7 billion years, as the universe is expanding, we gradually
drift away from the center of the Big Bang, and these frequencies are redshifted to
the microwave band. The cosmic background measured by the two engineers is the
remnant of the Big Bang. During the Second World War, due to the advent of higher
frequency radar, scientists gradually found that the Sun and some other celestial
bodies also have strong electromagnetic radiation.

Starting in the 1950s, Martin Ryle (1918–1984) and Antony Hewish (1924) of the
Cavendish Laboratory at the University of Cambridge of the United Kingdom used

interferometry method to produce high-resolution radio images of the universe and they also made a radio survey of the sky.

The four major astronomical discoveries of the mid-1960s are quasars, pulsars, interstellar molecules, and cosmic microwave background radiation. All these discoveries were achieved by radio astronomy. Among them, cosmic microwave background radiation is a milestone discovery in radio astronomy, which provides convincing evidence for the Big Bang theory. For this reason, the discoverers of pulsars and cosmic microwave background radiation won the Nobel Prize in physics in 1978.

2.4   A Brief History of Human's Access to Space

Many fascinating stories about flying to space are recorded in ancient Chinese books. The ancient Chinese also invented gunpowder. There is a long development history for ancient Chinese rockets, which were introduced to the West through Arabia and India around the middle of the thirteenth century. Around 1500 AD (Ming Dynasty), an inventor called Wan Hu[3] (real name: Tao Chengdao) put his flying dream into practice. He designed a wickerwork chair with 47 rockets underneath for liftoff and held two kites above, taking advantage of the rocket propulsion and the kites for a soft landing. He ordered his co-worker to ignite the rockets and tried to fly into space, but ended up catastrophically. This event shows that the Chinese has a long tradition of imagining flying into space and are keen to make efforts to realize the dream. A crater on the moon was named "Wan Hu" during an International Astronomical Union (IAU) meeting in the 1970s. After the Opium War, the awakened Chinese, as represented by Lin Zexu, advocated the introduction of Western technology to improve China's weapons and machinery. With this setting, Chinese began to develop modern rockets in 1851.

For the modern science period, the following scientists and engineers have made substantial and prominent contributions to human effort of gaining access to space.

Konstantin Tsiolkovsky (1857–1935) (as shown in Fig. 2.11) is a Russian scientist known as the *father of spaceflight*. He was the first to study the possibility of using rockets for interplanetary travels, as well as the possibility to make artificial satellite and build near Earth orbital station. He pointed out the reasonable way to develop astronautics and build rockets, and proposed a batch of important engineering solutions to rocket development and liquid engine structures. One well-known saying from Konstantin Tsiolkovsky is that "the Earth is the cradle of humanity, but mankind cannot stay in the cradle forever".

Robert Hutchings Goddard (1882–1945) (as shown in Fig. 2.12) is the maker of the first liquid rocket in the United States and the first person in human history to use liquid fuel for spaceflight. From 1926 to 1942, Goddard and his team launched

---

[3] In fact, Wanhu is a title for an official at that time.

Fig. 2.11 Sketch of
Konstantin Tsiolkovsky

Fig. 2.12 Sketch of Robert
Hutchings Goddard

34 rockets, and the most powerful can fly as high as 2600 m. Later, he led the team and conducted numerous experimental flights during the World War II. Although the high altitude required to enter space was not yet to be reached, the team set up plenty of technical standards for liquid rocket engines and applied for many patents, which laid the foundation for liquid rocket experiments in the United States. Later, the National Aeronautics and Space Administration (NASA) named its Maryland Flight Center after Goddard.

Frank Malina (1912–1981) (as shown in Fig. 2.13) and **Qian Xuesen** (1911–2009) (as shown in Fig. 2.14) are both recipients of America's first Scientific Achievement Award, and they are students of Theodore von Karman (1881–1963), a great aerodynamics master and Professor of California Institute of Technology (Caltech). In the early 1930s, Malina started the work to develop liquid rockets and set up a rocket team. The team members include mechanical engineers and chemical engineers, but the team needs an expert in theoretical analysis. Therefore, he invited Qian Xuesen, another student of von Karman, to join in for the calculation of orbit and solving the aerodynamic problems [3]. During the World War II, learning that Germany was speeding up its development of rockets, Qian Xuesen made a proposal to the US government in the name of von Karman for the establishment of the Jet Propulsion Laboratory (JPL). Therefore, the name of JPL first appeared in Qian Xuesen's proposal. For this reason, Qian Xuesen is regarded as one of the co-founders of JPL. The liquid rocket engine they developed was the most advanced in the United States

Fig. 2.13 Sketch of Frank
Malina

Fig. 2.14 Qian Xuesen

at that time, and the flight altitude records were constantly broke by the team, leaving
the altitude records of Goddard's rocket far behind.

Wernher von Braun (1912–1977) (as shown in Fig. 2.15) is the most outstanding
scientist and engineer in the field of modern rocketry. In the 1930s, with great interest,

Fig. 2.15 Sketch of
Wernher von Braun

he joined the German rocket group and later participated in the development of guided missiles. The missile in question is the most advanced guided weapon at the time which was later named V2. Before the end of the World War II, a large number of missiles were fired from Germany with London as the target, causing great casualties and public panic. It was the nightmare for the Allies. Before the allied occupation of the V2 rocket development base and the production base, von Braun fled to southern Germany with a large number of senior engineers and technicians, and surrendered to the American army. Some of them were transported to the United States, while some remained in France. Those who went to the United States became the core research team for the rocket of the first American satellite, while those who stayed in France became the core technical team for the development of the Ariane rockets. After the end of World War II, Qian Xuesen, dressed in American military uniform, accompanied von Karman to Germany to investigate the technology of V2 rocket and participated in the interrogation of von Braun. After that, the United States shipped home several unused V2 rockets.

As soon as the Soviet Red Army occupied Berlin, they immediately sent troops to capture the V2 rocket development sites, only to learn that von Braun and his technical team had fled. The Soviet army called back some skilled technicians to continue the production of V2 rocket engines. After the production capacity was restored, the factory was totally relocated to the Soviet Union.

The development history of V2 rocket and its inheritance relationships with rockets of various countries are shown in Fig. 2.16.

The Soviet Union launched the first man-made satellite Sputnik-1 (Fig. 2.17),

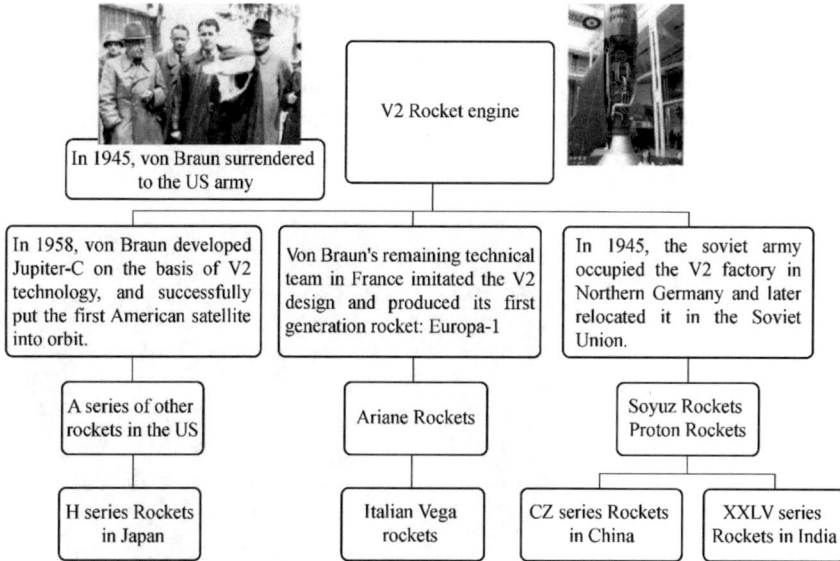

Fig. 2.16 Evolution of the German V2 rocket and its inheritance relationships with other countries' rockets

Fig. 2.17 Sputnik-1, the
first artificial satellite

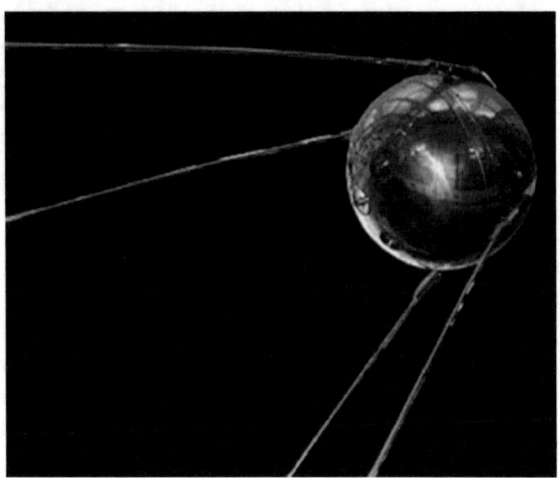

marking the beginning of the space age and the beginning of the space race between
the United States and the Soviet Union. In fact, before that, both sides had been
working,independently, on satellites programs and these activities were top secrets.
The American's development progress was hampered by several embarrassing failures, mainly because the American military, which was in charge of the development,
did not trust von Braun, who had surrendered from Germany, and relied solely on
the United States Navy to develop the launch vehicle for the first satellite. After the
launch of the first satellite in human history by the Soviet Union on October 4, 1957,
von Braun actively proposed to the U.S. Department of Defense and assured the latter
that a satellite could be launched within 3 months. Then, Braun got support from the
U.S. Department of Defense. He kept his promise 3 months later. The United States
successfully launched its first artificial satellite, Explorer I (Fig. 2.18) on January 31,
1958.

Fig. 2.18 Explorer-1, the
first American artificial
satellite. Photo credit: NASA

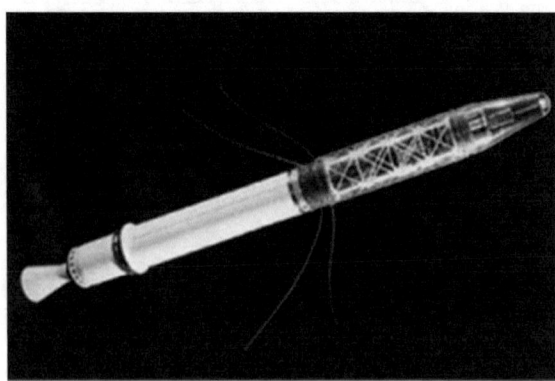

Shortly after that, the United States and the Soviet Union focused on the competition of "who is No. 1", such as the first manned spaceflight [the Soviet Union's Yuri Alekseyevich Gagarin (1934–1968), Fig. 2.19], the first extravehicular activity, the first female astronaut, the first manned lunar orbiting, the first manned moon landing [Neil Alden Armstrong (1930–2012) became the first man landed on the moon, Fig. 2.20], the first manned space laboratory, the first manned space station, etc. In general, the Soviet Union got the upper hand in the initial stage of the space race, making many records. But the United States caught up and gradually overtook the Soviet Union.

At the later period of the Cold War, the United States and the Soviet Union began to conduct joint space research, such as the docking of Soyuz with the Apollo. But the space race did not end until the dissolution of the Soviet Union in the early 1990s. It costs humongous amount of human and financial resources for both sides, but also greatly accelerated the pace of human's exploitation and utilization of space.

In 1990s, the United States became the only space power, and its focus gradually shifted to space science since then. Several important space science missions have been launched into space, including the Hubble Space Telescope (HST).

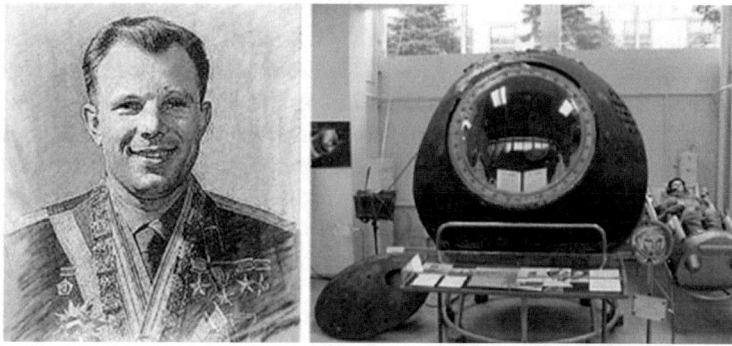

Fig. 2.19 First manned spaceflight. Soviet Union's Yuri Alekseyevich Gagarin became the first man entering space

Fig. 2.20 First moon landing: Neil Armstrong became the first man landed on the moon with the Apollo 11 mission. On the far right is the shoe print left by Buzz Aldrin (1930). Photo credit: NASA

Fig. 2.21   Zhao Jiuzhang

After the founding of the People's Republic of China, Qian Xuesen returned to China in 1955 after many setbacks. He immediately made proposals to develop rocketry, missiles, and the follow-up aerospace technology, and the proposals covered management, research, design, and production.

In 1958, China obtained the prototype of P2 missile engine from the Soviet Union, which was an improvement on the German V2 rocket. The project to imitate the P2 missile is called *Project 1059*. In 1964, China successfully tested a rocket capable of launching satellites, which is based on the work of imitating the engine provided by the Soviet Union. In 1965, China restarted the development of artificial satellites. China leveraged the development of ballistic missiles to study the launch vehicles, especially the liquid ballistic missiles, hence forming the CZ series launch vehicles. Since the Dongfanghong-1 (DFH-1),[4] Chinese space program started from scratch, and great leap forward was made from conventional propellants to cryogenic propellant, from multistage rockets to strap-on rockets, from one satellite a launch to multi-satellites a launch, from Low Earth Orbit (LEO), Sun Synchronous Orbit (SSO) to Geosynchronous Earth Orbit (GEO), from the launch of unmanned satellites to the launch of manned spacecraft, as well as lunar probes. In addition, China successfully entered into the market of the international commercial satellite launch service.

Zhao Jiuzhang (1907–1968, Fig. 2.21) is the founder of Chinese artificial satellite program. In 1933, he graduated from the physics department of Tsinghua University. In 1935, he went to Berlin University for further study and obtained his doctor degree in 1938. After returning to China, he was appointed the Director of the Institute of Meteorology, the Academia Sinica, and then the Director of the Institute of Geophysics, Chinese Academy of Sciences. In 1958, he was appointed the Deputy Director of the 581 Group which is responsible for the development of China's first artificial satellite. Through arduous effort, he established China's first research institute of space physics and the first ground simulation laboratory for space environment. In 1964, once again he put forward the proposal of developing China's

---

[4] *Note:* Dongfanghong means, literally, the East is red.

artificial satellite and formulated the development plan. In 1965, when the Chinese satellite program reopened again, he was appointed the Director of the 651 Design Academy. In 1968, when the Dongfanghong-1 (DFH-1) satellite entered engineering qualification phase, Zhao Jiuzhang passed away. Sun Jiadong took over the torch to develop DFH-1 satellite, which was successfully launched in 1971. In 1999, the Central Committee of CPC, the State Council, and the CMC posthumously awarded Zhao Jiuzhang the *Two Bombs and One Satellite Award*.

2.5   The Recent Technology Progress of Space Exploration

On October 4, 1957, the Soviet Union successfully launched the first artificial satellite, marking the beginning of space age, and since then the effort of space exploration never ceased. This section will briefly introduce the technical frontiers of launch vehicle, satellite and spacecraft, communication system of Tracking Telemetry and Control (TT&C), and launch and recovery technologies.

2.5.1   *Rocketry*

Up to now, the largest rocket ever built is NASA's Saturn V rocket in the 1960s. Saturn V looms a staggering height of 111 m (about the height of a building with 36 stories) and has a liftoff weight of 2800t. So far, the Saturn V rocket (shown in Fig. 2.22) is still the upper limit of human spaceflight.

Space shuttle, as a reusable manned spacecraft, can send satellites and spacecraft into orbit just like a launch vehicle. It can also operate in orbit like manned spaceship and can glide through the atmosphere for soft landing like a glider. The United States is the only country that has successfully completed manned missions with space shuttles, including Columbia (disintegrated on return), Challenger (exploded shortly after launch), Discovery, Atlantis, and Endeavour. Space shuttles retired in 2011.

SpaceX, an American space commercial company, has designed and built reusable commercial rocket Falcon 9, the first rocket that realizes controlled vertical landing on land and sea.

In the future, rocket propellant will be safer and more environmentally friendly, non-toxic, and pollution-free. Non-toxic and pollution-free liquid oxygen kerosene and liquid hydrogen oxygen propellant have been used in Chinese launch vehicles, such as CZ-5 and CZ-7.

Heavy rockets are necessary to enhance human being's capacity of gaining access to space beyond the Earth orbit. NASA will upgrade the Space Launch System (SLS) incrementally in three phases, with the aim of completing the Block 2 rocket with Low Earth Orbit capacity of 130t.

Fig. 2.22 Saturn V rocket

The advantages of solid rocket launch system include fast response and mobile launch. Besides the land-based launch, launch vehicles with solid propellant are also suitable for sea-based launch and airborne launch.

Access to space is expensive. As for international commercial LEO launches, the typical cost is $3,000–$5,000 per kilogram using small launchers (SpaceX). But the price is still too high for many applications. At present, the lowest price for entering space is 100,000 Chinese Yuan per kilogram (price offered by China). Once the rockets become reusable, the target price can be lowered to 20,000–30,000 Chinese Yuan per kilogram.

The major technical and capacity constraints to launch rockets into space are the mass and cost.

2.5.2 Satellite and Spacecraft

With the rapid development of space technology, satellites with better performance and versatile functions have empowered mankind with unprecedented space application capabilities. On the other hand, satellites become lighter, smaller, and more intelligent. Satellites can be mass produced, and the application efficiency of satellites has been constantly enhanced.

So far, the largest man-made satellite launched by a rocket is the Environment Satellite (Envisat, Fig. 2.23) from the European Space Agency (ESA). Its folded solar

Fig. 2.23 Envisat in testing. Photo credit: ESA

panels and radars reach a staggering height of 10 m. The Imager for Magnetopause-to-Aurora Global Exploration (IMAGE) has the longest foldable boom which can be extended to 504 m. As a type of smallest satellite, CubeSat has been standardized (shown in Fig. 2.24). It consists of one or multiple cubes of $10 \times 10 \times 10$ cm. There is a project that is still in the design phase called Breakthrough Starshot Program, in which Stephen Hawking has participated. The program is sponsored by the American Breakthrough Foundation to develop light sail spacecraft called Starchip, which will be accelerated by powerful lasers located on the moon to reach the velocity of 20% light speed (about 6×10^4 km/s). With that speed for interstellar mission, the spacecraft will reach and explore centauri α in about 20 years. It takes about 4 years for the Earth to receive the message transmitted back.

The frontiers of satellite technology vary greatly according to missions' requirements. Nevertheless, to develop smaller, smarter satellites has emerged as the new

Fig. 2.24 CubeSat

trend, especially the design of intelligent formation (i.e., the formation of a larger scale satellite cluster by connecting separate satellite groups).

In addition to all types of satellites in the Earth orbits, the spacecraft heading to the moon, the Mars, the Venus, and other celestial bodies in the solar system also indicate new directions for space technology. In addition to the scientific objectives, these missions also aim at the utilization of the resources in the solar system, as well as the identification of small Near Earth objects (NEOs) that may threaten the Earth and require corresponding defense strategies. The innovative technologies involved include interplanetary navigation, landing on and liftoff from extraterrestrial bodies, sample collection and sample return, extraterrestrial survey and subsurface detection, as well as in situ sample analysis and utilization.

2.5.3   Tracking Telemetry and Control (TT&C) and Communication

Tracking, Telemetry and Control (TT&C) technology refers to the comprehensive technologies to track, measure, and control the target spacecraft. Tracking Telemetry and Control system consists of control centers, tracking and telemetry stations, and specialized communication network. TT&C technology is moving to the direction of continuous orbit coverage, and the precision of orbit determination improves from meter to centimeter.

Communication technology mainly refers to the technology of transmitting large amounts of data from the spacecraft to the ground. Due to advances in laser communication technology, the satellite-to-ground data transmission rate from Low Earth Orbit will soon surpass the rate of bandwidth (Gbps).

At present, TT&C technology is expanding into the field of deep space exploration. Besides the speed, the main obstacle of deep space TT&C is the time delay caused by long distance. For the communication between the Earth and the moon, the delay will be as long as 1.3 s, while for the communication between the Mars and the Earth, the delay will be 3–20 min depending on the orbits of the Earth and Mars. Figure 2.25

Fig. 2.25 Pegasus radio telescope, Shanghai Astronomical Observatory (SHAO)

shows the 65 m Pegasus radio telescope of Shanghai Astronomical Observatory (SHAO), Chinese academy of sciences.

2.5.4 Launch and Recovery

In the future, space launch will be diversified. Fixed launch sites will remain the only option for manned missions, as well as for large and heavy launch vehicles, while sea-based and airborne launches are complementary and will have new opportunities.

For manned spaceflight, it is compulsory for the launch sites to have the capacity to serve the astronauts, including training, medical care, quarantine facilities, and related technical support. Therefore, countries with capacity to launch manned missions all have designated launch sites for such missions.

When it comes to the site selection and design of each launch site, we need to consider the safety issue drop zone of the first stage. The rotation direction of the Earth is from the west to the east. Therefore, in order to take advantage of the power brought by the rotation of the Earth, for the satellites to enter orbits with inclination, the launch downrange will head to the east so as to save propellant. Of course, polar orbit satellites are exceptional, and their downrange could head to the south or the north. Therefore, to ensure the safety in the drop zone, the launch sites are normally located on the east coast, which allows the remains of the first stage to fall down into open seas.

For the unmanned recoverable satellites, such as microgravity missions, the space life science experiment satellites, and manned missions, vast plains are typically

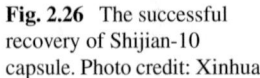

Fig. 2.26 The successful recovery of Shijian-10 capsule. Photo credit: Xinhua

preferable as the recovery sites. The reason behind is that in the recovery process, due to the uncertainty of meteorological conditions, after the parachute is opened, the real-time estimation of the landing point is very tricky. Therefore, the open prairie or desert areas are the best choice. However, in order to find the re-entry capsule to collect the samples and rescue the astronauts as soon as possible, it is necessary for the recovery sites to build corresponding technical facilities to meet the particular needs of the recovered cargos and astronauts. Figure 2.26 shows the successful recovery of Shijian-10 capsule.

References

1. Joseph N (1959) Science and civilization in China. Vol. 3: mathematics and the sciences of the heavens and the earth. Cambridge University Press, Cambridge
2. Joseph N (1991) Science and civilization in China. Vol. 2: history of scientific thought. Cambridge University Press, Cambridge
3. Chang I (1996) Thread of the silkworm. Basic Books, New York

Chapter 3
Major Frontier Issues in Space Science (I)

3.1 Introduction

Space science relies on spacecraft to carry out scientific research, with primary focus on the questions about space itself and the extended basic scientific questions regarding the universe. Its scientific frontiers can be boiled down to two major scientific questions: what is the origin of the universe and its evolution? and what is the impact of solar activities on human being?

The most ancient scientific question of human being involves the study of the universe and thoughts on the universe. Since the beginning of space age, basic knowledge of space environment and its characteristics has been accumulated and space physics studies have been conducted based on regional scientific data. After that, space astronomy spearheads the effort to use spacecraft to conduct relevant research. After nearly half a century, scientific breakthroughs achieved by spacecraft have substantially expanded our understanding of the universe. This chapter will introduce the universe we know in time dimension and spatial dimension, respectively, and then briefly introduce the questions of great significance in space astronomy.

Let us zoom in from the universe to the solar system where we live. Ever since human being gaining access to space, we began to explore and study space in order of proximity and distance. After the study of the ionosphere, the next stop is the Earth's magnetosphere, which is filled with charged particles and is controlled by the magnetic field of the Earth. Beyond the boundary of the magnetosphere, we study the interplanetary solar wind and with the help of remote sensing instruments, we observe the Sun's upper atmosphere all the way to the Sun's surface at various frequency bands. This chapter will follow an opposite route by unfolding the process of the impact of solar activities on human activities, from the basic knowledge of the Sun, all the way to the ionosphere and upper atmosphere, after which the relevant scientific frontiers will be discussed.

© Science Press 2021
J. Wu, *Introduction to Space Science*, Springer Aerospace Technology,
https://doi.org/10.1007/978-981-16-5751-1_3

3.2 Origin of the Universe and Its Evolution

3.2.1 Time Dimension

That the universe began with a Big Bang has been proven by modern scientific observations. From the moment of the Big Bang, the universe was born. In fact, for theoretical physics calculations, the Planck time is the shortest length of time, which is the interval of time from 0 to 10^{-43} s (known as Planck time) after the Big Bang. After that, the basic forms of matter began to emerge, including quarks, neutrinos, protons, electrons, and photons. Therefore, we normally regard Planck era as the starting point of the universe.

10^{-35} s after the Big Bang [1], the universe cooled to the point where strong force can separate, and inflation occurred, expanding the universe for 100 times. The resulting scale is 10^{30} times larger than the previous one.

Almost in the blink of an eye, after one millionth of a second since the birth of the universe, light appeared. In addition, the fundamental building blocks (protons, neutrons, and electrons) emerged which makes up the matter we observe in the universe today.

10 seconds after the Big Bang, stable nuclei (chemical elements) such as hydrogen and helium were formed, which ushered in the age of thermonuclear reaction.

Somewhere between a few hundred thousand years and 100 million years, there is plenty of space in the universe for the neutral hydrogen. But as the photons dwindled, the universe became very dark, which is known as the *"Dark Age" of the universe*. In 1945, a Dutch scientist accidentally discovered that neutral hydrogen also emits electromagnetic radiation in microwave frequency band. The physical mechanism was later confirmed as: when an electron revolves around the nucleus with a proton, it will occasionally change the spin direction, shifting from clockwise to counterclockwise, or the other way round. Once the change occurs, it emits a 1.4 GHz microwave pulse. If a probe is inserted into a closed waveguide filled with neutral hydrogen, a 1.4 GHz pulse can be detected from time to time. That little bit of information provides the basis for studying and observing the Dark Age of the universe. Of course, after the expansion of the universe for 13.7 billion years, the 1.4 GHz microwave radiation has redshifted to the shortwave frequency band of 1–30 MHz, making it impossible to observe within the Earth's ionosphere (Fig. 3.1).

For some reasons, neutral hydrogen is not evenly distributed in the primordial universe. Due to gravity, the neutral hydrogen collisions and compressions occurred, triggering ionization and nuclear fusion and hence light appears in the universe again. This stage is called *"dawn of the universe"*. And then stars formation began, a process taking roughly 1 billion years.

As the number of stars grows to some point, galaxies formed. Galaxies encompassed a large number of stars. Most galaxies are stable, though galactic cannibalism occasionally happened. The life span of stars within galaxies varies from a few billion years to 10 billion years, depending on their size. The Sun, for example, is only 4.6 billion years old and is still in its mid-age. Stars smaller than the Sun have a longer

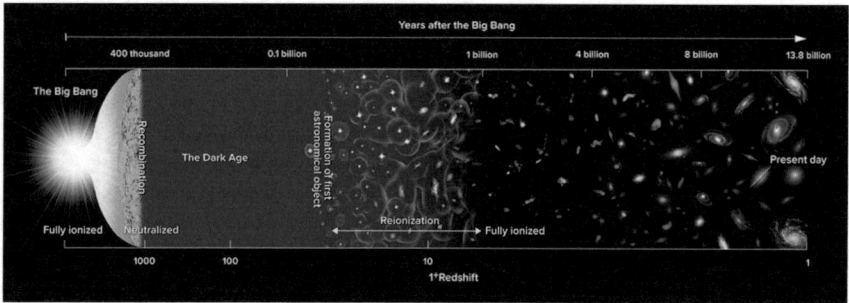

Fig. 3.1 A timeline of the expansion of the Universe. Image credit: NRAO

life span, while larger stars have a shorter life span. For those large enough, it will evolve into dense neutron stars or even black holes at the end of their lives. Our Sun belongs to the second or even third generation of stars in the Milky Way. In other words, the death of the previous stars produced immense interstellar clouds, which concentrated under gravity, hence forming the Sun. The ongoing mission of Hubble Space Telescope can observe many first-generation stars in the most distant reaches of the universe.

According to the observations of distant universe by the Hubble Space Telescope, the age of the universe is around 13.7 billion years old. According to the redshift data, it also takes 13.7 billion years for the light from the most distant galaxies to reach us. Redshift refers to the phenomenon that the electromagnetic radiation of a celestial body moves toward longer (red) wavelengths as it gradually moves away from us. In the visible band, it shows that the spectral line of the spectrum shifted somehow toward the red end, and hence the wavelength becomes longer and the frequency becomes lower. If the object we are observing is redshifted, it's receding from us. If we look deeper into space, we can see a broad spectrum of microwave radiation. The microwave background is the "sound" relics from the Big Bang after so many years of redshift.

The Earth we live on is almost at the same age as the 4.6-billion-year-old Sun. Life appeared on the Earth very long time ago and dinosaurs died out 65 million years ago. However, humans appeared on the Earth only 2 million years ago. It is even noteworthy that human being enters into the age of civilization with writing systems only in the past 5,000 years, and humans enter modern civilization only in the past centuries.

As you can see, in the time dimension of the universe, if we regard the period from the birth of the universe to the present as a day, the real civilization of human beings lasts probably less than a second, in the blink of an eye. In such a short period of time, bit by bit, humans have been able to trace back and reveal the long history of the universe. This makes science, and space science in particular, so fascinating.

3.2.2   Spatial Dimension

Three units of distance are introduced as follows to explain the spatial dimension of the universe.

The first is the Astronomical Unit (AU). 1 AU is equivalent to the average distance from the Earth to the Sun, about 150 million km. AU is an ideal unit to describe the distance in the solar system. For example, the distance between the Mercury and the Sun is 0.39 AU, from the Venus to the Sun 0.72 AU, from the Mars to the Sun 1.52 AU, from the Jupiter to the Sun 5.2 AU, from Saturn to the Sun 9.54 AU, from the Uranus to the Sun 19.19 AU, from the Neptune to the Sun 30.07 AU, and from the solar wind Termination Shock to the Sun 130–150 AU. Voyager I, the human spacecraft that travels the farthest, has now reached the distance of 150 AU. The space science community defines the Termination Shock as the boundary of the solar system. The solar wind "blows" outward, and when it "blows" to a certain point, it interacts with the medium between the stars, forming a sort of "wall" called Termination Shock. Therefore, the space physics community believes that Voyager 1 has already left the solar system. However, the astronomy community thinks differently and they believe that the boundary of solar system is defined by the distance to Kuiper Belt, about 30 AU.

The second unit is light year, or the distance that light travels in the course of 1 year. A light year is about 9.46 trillion km (63,000 AU). The nearest star to our solar system is Proxima Centauri, which is about 4.2 light years away. The Sun is located on one of the spiral arms of the Milky Way, about 26,300 light years from the center of the Milky Way. Figure 3.2 shows the position of the Sun in the Milky Way. The diameter of the Milky Way is about 100,000 light year and there are 100 billion to 200 billion stars in the Milky Way. The light year as a unit is used to describe the Milky Way and other distant objects.

The Milky Way is just one of hundreds of billions of galaxies in the universe. The nearest galaxy to the Milky Way is Andromeda, which is about 2.2 million light years away. It is 200,000 light years in diameter, larger than the Milky Way. Between the Milky Way and Andromeda, there are some scattered clusters of intergalactic matter called dwarf galaxies.

Looking deeper into the universe through the Milky Way's stars, the Hubble Space Telescope's 0.5° field of view (FoV) can observe about 100 galaxies. In general, the galaxies in the universe are evenly distributed. By this reckoning, there are more than a billion galaxies or even up to 100 billion in the observable universe. The Milky Way is a medium-sized spiral galaxy, there are other galaxies of different shapes in the universe, such as the elliptical galaxies.

The third unit is the parsec (pc),[1] which represents the distance at which the radius of Earth's orbit subtends an angle of one second of arc. Thus, a star at a distance of one parsec would have a parallax of one second. Apparently, parsec is a much larger unit of astronomical distance. Besides the light years, parsec is often used to describe the distance from a star to the solar system.

---

[1] 1 pc ≈ 3.26 light-years.

Fig. 3.2 Our solar system's place in the Milky Way. Image credit: space.com

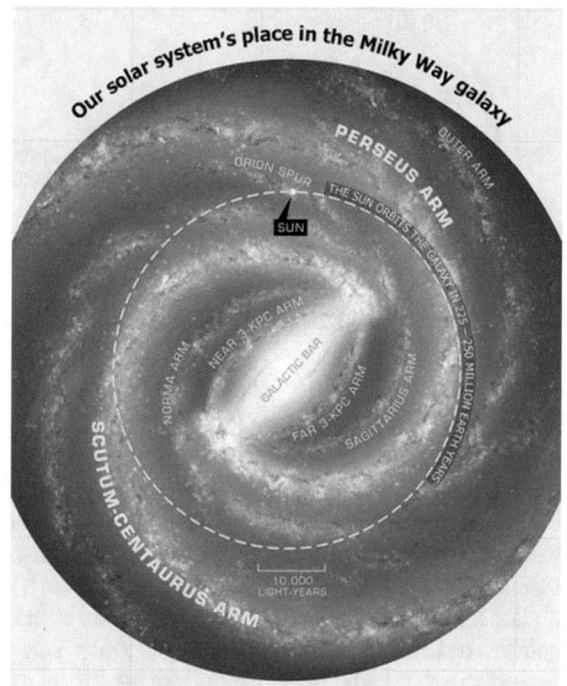

3.2.3   Questions of Great Significance

(1) The Origin, Evolution, and Future of the Universe

According to observations, the universe is now expanding at an accelerating rate. This is quite puzzling. If the universe originates from a Big Bang, then there would be periods of accelerated expansion, decelerated expansion, equilibrium stagnation (silent universe), and after that a period of contraction as explained by the theory of gravity, and eventually the universe will return to the point where the Big Bang occurred.

We are now in a period of accelerated expansion. Where does the energy for anti-gravitational expansion come from? What is the energy driving the Big Bang? Is it still working? Is the universe, aging more than 13 billion years, still in its infancy? Will the expansion decelerate and come to a stop? How long will the stagnation last? Therefore, the origin, the evolution, and the future of the universe are important scientific questions.

(2) Dark Matter and Dark Energy

As shown in Fig. 3.3, the ordinary matter accounts for only about 5% of the universe, with the rest being dark matter and dark energy. The theory of gravity provides the basis for the existence of dark matter. If the theory of gravity can be applied

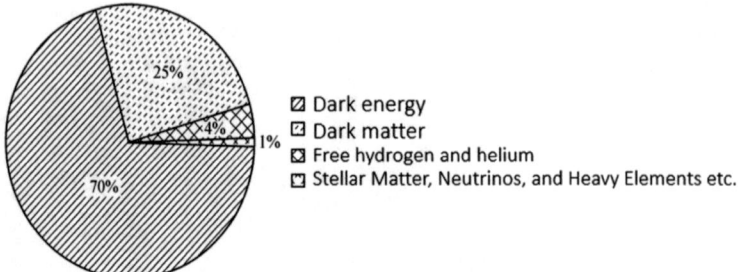

Fig. 3.3 The basic matter composition of the universe as jointly defined by the current astronomical data

universally, dark matter could help explain why the Sun spins so fast, circling the center of the Milky Way. Such a rapid speed would generate a centrifugal force five times greater than the binding force of gravitation generated by all the ordinary matter combined in the Milky Way. The viable explanation is that in the Milky Way, there exists dark matter five times heavier than the ordinary matter, which is the source behind such a strong gravitational pull.

Observations of the orbiting speeds of stars around other galaxies reveal that all galaxies have the same phenomena as the Milky Way. Theoretical physicists are also exploring new theories to modify current theories of gravity. If the theories of Newton and Einstein concerning gravity are not correct when it comes to a large scale, then it is possible that dark matter does not exist. That is to say the ordinary matter is sufficient enough to generate such a strong gravitational pull to constrain the rotation of stars. But recent observations have found that stars around certain individual galaxies don't rotate as fast as other galaxies, which matches the gravitational pull generated by the ordinary matter. This case proves, in a way, that the theory of gravity can be applied to large scale, and that there is no dark matter in those specific galaxies. But in galaxies with dark matter, when we scale down, e.g., to the solar system, there is no dark matter. The speed at which the eight planets orbit the Sun fits perfectly with the theory of gravity without dark matter.

If dark matter is also composed of elementary particles, one theory holds that dark matter is neutrinos, because as the most elementary particles known so far, neutrinos do not interact with other matter, or just with slight interactions. This has led to the idea of digging deep tunnels in the mountains where neutrino detectors could be placed to shield as much as possible the background radiation of other particles. It's a bit like sitting back and waiting to see if you can catch the neutrinos that make their way through the thick layers of the mountains to the detectors. But so far, nothing has been found.

China's Dark Matter Particle Explorer (DAMPE) mission, or Wukong (named after the legendary monkey king in the Chinese classic literature *the Journey to the West*), is hunting for the evidence of dark matter from another perspective. If dark matter is made up of particles, then the particles will collide with each other. According to analysis of theoretical model, the collision will result in ordinary matter particles with extremely high energy, such as electrons. This energy spectrum is around 10^{12} electron volts (1 TeV). The designed energy spectrum for the DAMPE mission is within 10 GeV–10 TeV. Based on the current observation data, anomalies in the energy spectrum do exist around 1.4 TeV.

Dark energy is accountable for the accelerated expansion of the universe. The only reliable way to detect dark energy is to accurately measure the positions of galaxies and stars in the universe, and then compare their positions over a period of time to see quantitatively how much the universe is expanding and whether the expansion is isotropic.

(3) Black Holes and "White Holes"

The existence of black holes as a celestial body has been confirmed. Large stars (several times larger than the Sun) collapse from within at the end of their lifetimes, hence forming black holes. Every galaxy tends to have a giant black hole at its center, and so does the Milky Way. The massive gravity of the black hole constantly pulls in the matter around it, gobbling up the massive objects, e.g., the stars, and dragging it out of shape which will eventually disappear into the black hole. Prior to falling into prey of black holes, ordinary matter is destined to undergo extreme physical conditions of extreme heat, pressure, magnetic field, and density. Under these conditions, lower energy radiation cannot escape from the gravitational pull of the black hole, while particles with higher energy can still break free. In other words, the boundary of the black hole gradually shrinks as the energy of the particles increases. If the telescope has sufficient spatial resolution, the higher the energy spectrum it can observe, the more visible the region close to the center of the black hole. Therefore, the various phenomena occurring around the black hole can be detected.

At present, theoretical physicists still cannot explain whereabouts of the matter and energy that fell into a black hole. This resembles entering an unfathomable pit. For mathematical models, the multi-dimensional universe models began to emerge, in which curved space would possibly create a "white hole" under the black hole and the "white hole" will "spit out" the matter and energy engulfed by the black hole. In recent years, the observation shows that active galactic nuclei (AGN) radiated out vast amount of energy. Although they are very far away from us, we can still receive cosmic ray radiation with extremely high energies from their directions. However, compared with that of black holes, the theory of "white holes" and the according observation is still very immature.

(4) Gravitational Wave

In 2016, the gravitational wave predicted by Albert Einstein (1879–1955) was detected for the first time by the ground-based Laser Interferometer Gravitational-Wave Observatory (LIGO), a 4-kilometer-long, right-angle-shaped laser interferometer. This event reaffirmed the correctness and universality of the interconnected space-time model of the universe as described in the theory of general relativity. The abrupt change in space-time will generate ripples, or gravitational waves which travel at the speed of light. Therefore, the events that cause matter distribution changes, such

Fig. 3.4 LIGO measurement signals in synchrony indicating that gravitational waves were first detected by humans. Credit: Science

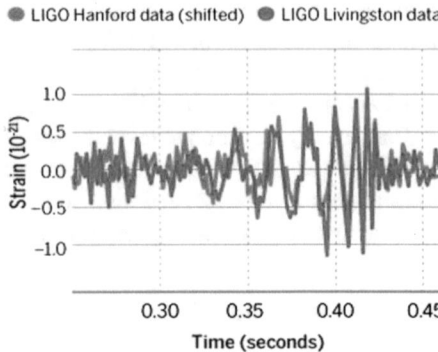

as the Big Bang, the merging of galaxies, stars, black holes, and the engulfing of stars by black holes, all generate gravitational waves. Taking into account the vast space and long history of the universe, such events must have occurred many times and the gravitational waves must have propagated through the vast universe over long periods of time. It is conceivable that the space-time of our solar system must be filled with gravitational ripples at different frequencies and directions, which are superimposed on each other and are changing over time. The gravitational wave pulses measured by LIGO since 2016 are only a few strong ones with higher frequencies. Figure 3.4 shows the first detection of gravitational waves signal using LIGO. In August 2017, several ground-based and space-based observatories around the world participated in a joint event observation, and most of them discovered the electromagnetic waves (from radio waves to X-rays) that were generated at the time of the event. We are justified to say that the era of multi-messenger (not only via the electromagnetic spectrum) observation of the universe is coming.

The baseline of ground-based laser interferometer is too short and only high-frequency gravitational waves can be observed. Therefore, it is necessary to venture into space for long-baseline gravitational waves observation, which will lead the direction of gravitational waves detection in the future.

(5) Exoplanets and Earth-Like Planets

Before 1995, we literally know nothing about exoplanets, because they had never been discovered. At that time, the solar system that has nine planets[2] seems very special. This unknown state about exoplanets was shattered by the discoveries from the Geneva Observatory in Switzerland. By measuring periodic variation of a star's spectrum, they confirmed that there must be planets orbiting the very star, and that the mutual gravitational pull among the star and its planets caused periodical slight variation in distance to us. These slight variations could be captured by precise spectral shift (Doppler effect). This detection method became known as radial velocity method.

---

[2] At that time, Pluto was still listed as a planet.

Since then, thousands of exoplanets have been discovered by space-borne and ground-based telescopes, and now it is known that almost every star is surrounded by a planetary system. For example, Proxima Centauri, the closest star to our Sun, has planets revolving around it.

As illustrated by the solar system, not all planets are suitable to harbor and nurture life, which gives rise to the term habitable zone. A habitable zone means that the planet is at a moderate distance from its star to keep its surface temperature between 0 and 100 °C, allowing liquid water to exist on its surface.

Exoplanets have been discovered, and now we understand that there are many exoplanets in the universe. However, the most intriguing and tempting research is the search for Earth-like exoplanets or the Earth's sister planets. Besides the existence of liquid water, there are several additional conditions for Earth-like exoplanets: (1) the host star should resemble the Sun. It should be a main sequence star, not a star with lower temperature such as a red dwarf; (2) a proper distance to the host star, i.e., it must be in the habitable zone (Fig. 3.5), where water can exist in liquid form; (3) it must be a planet with a rocky surface where intelligent life can live and prosper; (4) its size or gravitation should be at the same level as the Earth. That is to say, there is more or less the same gravitational pull on its surface as the Earth's gravitational field, so that life on the planet cannot grow too high (in case its gravity is much weaker than the Earth) or too short (in case its gravity is much stronger than the Earth). (5) It must have a magnetic field that shields it from energetic particles that come from stars and the universe; (6) it must have an atmosphere with a moderate proportion of oxygen; and (7) it must have a spin cycle similar to that of the Earth, rather than that of Venus which takes hundreds of days. For long spin cycles, there is even risk that the planet is tidally locked to the star, with only one side facing the

Fig. 3.5 Habitable zone. Image credit: NASA

star, forming a stable near side and far side. If these conditions are met, we can call it an Earth-like planet. So far, no such planet has been discovered.

Currently, the United States is accelerating the pace of Exoplanet exploration. The Kepler mission is followed by the launch of the Transiting Exoplanet Survey Satellite (TESS) mission which focuses on the search of exoplanets close to the solar system. In addition, the search of exoplanets is listed as one of the main scientific objectives of James Webb Space Telescope (JWST). Chinese scientists have proposed a project called Close-by Habitable Exoplanet Survey (CHES), aiming directly at the search of Earth-like planets instead of exoplanets.

So the question is, is there such a planet in the universe? If so, are there any intelligent life on those planets? This is how the planetary science and space life sciences begin to intersect.

3.3   The Impact of Solar Activities on Human Being

3.3.1   Solar Activity

The Sun keeps a spin cycle of 27 days from the early stages of its formation. The rotation direction of the Sun follows the right-hand rule with thumb pointing to the North Pole of the solar system and other fingers indicating the direction for the Sun and the planets. It is the same with the planets in the solar system. The Sun has an 11-year activity cycle, which is accountable for the periodical changes of sunspot numbers and other phenomena.

The Sun has a magnetic field and is constantly ejecting particles into the interplanetary space, creating a steady solar wind. The particles are mainly composed of ionized hydrogen (i.e., protons and electrons). In the absence of intense activity, the speed of solar wind is 350–400 km/s.

The temperature of the Sun's interior is up to 10 million degrees, where nuclear fusion has been taking place over a long period of time, producing and transporting large quantities of photons to the surface. The temperature at the Sun's surface is only about 6000 °C. Intense activities take place on the surface of the Sun, and the resulting cryogenic zone is called a sunspot. China has the earliest record of sunspots, dating from the Western Han Dynasty (28 BC), almost 1000 years earlier than that written in the Western literature. About 200 years ago, European scientists began to record the number of sunspots to define the Solar Maximum Year. We are currently entering the 25th solar cycle.

The Sun's surface undergoes dramatic changes in the electromagnetic field, emitting large quantities of photons and electromagnetic waves with broad spectrum. This phenomenon is called "flares", which exerts direct influence on the Earth's space environment. Driven by the electromagnetic field and the unbalanced energy, the Sun also intensively ejects masses, known as Coronal Mass Ejection (CME). CME is shown in Fig. 3.6. The density of CME ejected into the interplanetary space

Fig. 3.6 Coronal Mass
Ejection (CME). Photo
credit: ESA/NASA

is very high and the magnetic field of CME is much stronger than the solar wind, resulting in Interplanetary Coronal Mass Ejection (ICME). The velocity of ICME can reach 800 ~1200 km/s. CME was only discovered after the launch of man-made satellites and we began to realize that CME and ICME are the most important factors that exert influence on the Earth's space environment.

3.3.2   Interplanetary Space Weather

The ICME superimposed on the solar wind travels very fast, with velocity exceeding 1000 km/s. The ICME propagates into the interplanetary space in the form of rotating ejections. If multiple CMEs occur sequentially and they have different interplanetary velocities, they will recombine to form new distribution in the interplanetary space. Since ICMEs are regional, whether they will reach the Earth is not certain. There is possibility that they go in the opposite direction or just fly by the Earth. Therefore, whether an ICME can evolve into space weather events in geospace depends on the location where it occurs and on its propagation path in interplanetary space. In fact, only about 50% of CMEs observed on the Earth (i.e., occurring on the near side of the Sun) can reach the Earth and consequently form space weather events.

Spacecraft flying in the interplanetary space is also subject to ICMEs. When they encounter, the safety of the spacecraft is at stake. The man-made probes, landers, etc. sent to the Venus, the Mars, and various small celestial bodies can also be affected by solar eruptions. In a word, the concept of "space weather" has extended from the geospace to the entire solar system.

3.3.3   The Magnetosphere of the Earth

Except the Venus and the Mars, the rest of the planets in the solar system all have magnetic fields. Among them, the Mercury, the Earth, and the Jupiter have similar magnetosphere, whose magnetic fields are dipole magnetic fields with north and south poles. Such a magnetic field interacts with the solar wind, creating a magnetosphere that protects the planet. A large number of particles (usually ionized charged particles) are blocked by the magnetic shell and then move along the magnetic field lines toward the poles. For those particles entering the "funnel" of the polar cusp region, frictions occur with particles in the atmosphere (for planets with atmosphere), hence producing aurora. If pilots, who fly the polar region, regularly experience space weather events, they will be exposed to increasing particle radiation, which affects their health. Figure 3.7 is a simulation of the Earth's magnetosphere.

Be it the Mercury, the Earth, or the Jupiter, planets with dipole fields have their own radiation belts. Protons are accelerated in the inner radiation belts, forming high-energy protons, while electrons are accelerated in the outer radiation belts, forming high-energy electrons. These two regions are the focus of space weather research. In particular, due to the weakening of the Earth's own magnetic field over the South Atlantic, a magnetic field anomaly has been formed, and the altitude of inner radiation belts above this region dips down to several hundred kilometers, posing a threat to

Fig. 3.7 A simulation model of the Earth's magnetosphere

the satellites on the Near Earth Orbit. This is called the South Atlantic Anomaly (SAA).

When space weather events occur, the Earth's magnetosphere is subject to violent turbulence, called magnetic storms. The magnitude of magnetic field can rise and fall by up to 20%. The year 1859 witnessed a major solar eruption, causing tremendous changes in Earth's space weather, known as the *Carrington Event*. At that time, the most advanced technological infrastructure on the ground was the wire telegraph lines used for communications. Changes in the magnetic field led to an induced current in long cables, which was discharged at the terminals, setting fire on the telegraphs. It has been documented that in Mexico City, at a magnetic latitude of 30–40°, people could read a newspaper at night under the aurora, indicating the brightness of the aurora. And in 1989, a solar eruption damaged electric power grid near the polar region in Canada's Quebec Province, leaving millions of people in blackout [2]. This is also due to the strong induced currents entering the electric power lines. This space weather event also caused damaging disruptions to satellites in orbit, with several satellites damaged or even completely disabled by dramatic increase in high-energy particle radiation. Magnetic storms, however, occur with time delay. Therefore, precautions can be taken by early warning before the solar wind is about to reach the Earth's magnetosphere.

In contrast to magnetic storms, the frequently happened small disturbances of the magnetosphere are called substorms. It is currently believed that the source region of substorms is the magnetotail, a long, open region of the Earth's magnetic lines at the night side of the Earth. The magnetic field in the magnetotail is very unstable, and once reconnection occurs, it reverses and transports the charged particles back to the Earth, forming a substorm. Substorms can also cause, to some degree, space weather events, but they do not significantly affect ground infrastructures.

3.3.4 The Earth's Ionosphere

The ionosphere is created when particles in the upper atmosphere are excited by ultraviolet light from the Sun and electrons escape from the confines of the atoms. During the day, the lower boundary of ionosphere is as low as about 60 km. This region is called the D layer which disappears at night, and the electrons automatically combine with atmospheric ions to form neutral atmosphere. The E layer, extending from an altitude of 90–110 km, is the stable layer of ionosphere, which does not completely recombine even at night. The F1 (200 km) and F2 (300–400 km) layers are stable regions further up. The reflection of radio waves in the ionosphere is shown in Fig. 3.8. However, when space weather events occur, the ionosphere will experience major fluctuations, which is called an ionospheric storm.

Fluctuations in the ionosphere affect the radio signal passing through it, causing changes in the signal curvature. As a result, a shortwave signal received in the condition of stable ionosphere will disappear when an ionospheric storm occurs. Global Positioning System (GPS) signals that pass through the ionosphere to reach

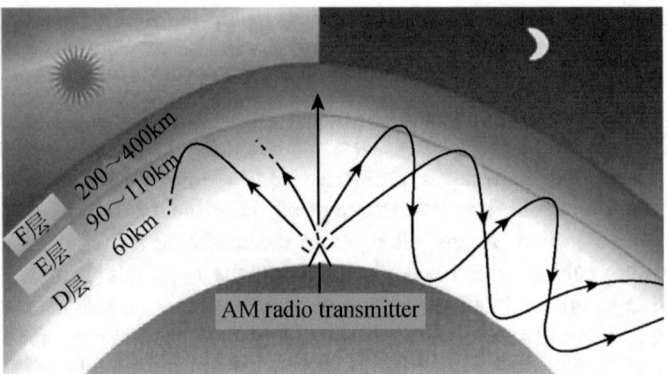

Fig. 3.8 Radio wave reflection in the ionosphere

the ground may also change, leading to the inaccurate positioning. Currently, it is common to use double frequency positional GPS signals for correction, or to use local ionospheric measurements to correct navigation signals in real time in a bid to improve positioning accuracy. Therefore, real-time ionospheric detection data has certain significance in terms of safety concerns and strategic values.

At low latitudes, the E region of ionosphere also contains large, shifting irregular regions (Sporadic E clouds), ranging from a few kilometers to tens of kilometers in scale, drifting from west to east. These inhomogeneous structures can also bring about rapid changes in radio signals, which is called scintillations. This is also a major focus of space weather research in the ionosphere.

During a space weather event, ionospheric storms can occur in the ionosphere, but they are delayed roughly by a few days. In other words, once a major space weather event occurs, it is feasible to make a short-range forecast of ionospheric storms.

3.3.5   The Middle and Upper Atmosphere

The middle and upper atmosphere, as shown in Fig. 3.9, is a region dominated by the neutral atmosphere 20–90 km to the Earth surface, which is the main flight area for near space spacecraft. The atmosphere in this region, though thin, is highly unstable and vulnerable to disturbance. The sources of disturbance come downward from the troposphere under it, as well as upward from the Sun and the upper ionosphere.

In the past 10 years, more attention was paid to the research of middle and upper atmosphere. There are two main reasons accountable for this shift. First, the near space application begins to be valued. Although the atmosphere in this layer is thin, if the velocity of the aircraft is very high, it can still gain some lift, which explains the emergence of extreme high speed, winged space vehicle. However, the considerable fluctuations in the middle and upper atmosphere will have great impact on the lift

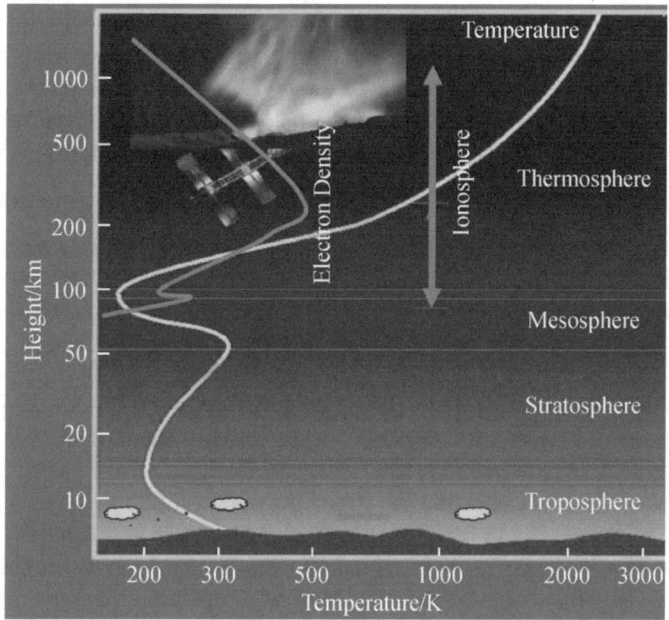

Fig. 3.9 The middle and upper atmosphere

force of the aircraft flying at high speed. Second, its relationship with space weather gradually became one of the focuses of the study of global change. The middle and upper atmosphere is regarded as the transition zone and medium between the weather and climate of troposphere, and the space weather and climate.

The space weather phenomena in the middle and upper atmosphere are mainly related to temperature and density fluctuation, as well as composition changes. In addition, there are transient phenomena such as *red spirit* and *blue jet*. With the advent of satellites, the *red spirit* and *blue jet* were gradually observed, which the Americans initially mistook as Soviet missiles, only to find later that they were natural phenomena.

3.3.6 Questions of Great Significance

(1) The Solar Eruptions and Its Forecast

Solar eruptions are the origins of space weather events. However, our ability to predict solar eruptions is still weak. Predicting flooding by counting sunspots, a sort of fortune-telling, has long been thought to have no scientific basis. Though we have good observations of flares and CMEs, space weather cannot be well predicted. In recent years, with the improvement of our ability to observe the Sun, we have been gradually paying attention to the changes of solar magnetic field. These changes are considered as important physical parameters that may reveal the causes of flares and CMEs. The Advanced Space-Borne Solar Observatory (ASO-S), a Chinese scientific mission, aims to observe the solar magnetic field, flares, and CMEs simultaneously to study the relationships among them and the corresponding patterns.

(2) Interplanetary Solar Wind and Its Interaction with the Magnetosphere

The propagation path of CME into interplanetary space is also a factor that exerts direct influence on space weather. Wherever ICME goes, the magnetic field fluctuates, and the energy and density of particles increase sharply, which brings harm to the interplanetary space activities. The ICME's encountering with the Earth or planetary magnetosphere will bring about space weather events for the Earth and other planets. For the Earth, the changes from the magnetopause to the poles and from the magnetic field to ionospheric storms, can affect man-made satellites and large ground infrastructures over a period of several days.

Because of the Sun's spin and the strong gravitational pull in the ecliptic plane, even the CME occurring in the high latitude of the Sun will gradually diffuse in the ecliptic plane after entering the interplanetary space. However, due to the changes in the initial velocity and direction when leaving the surface of the Sun, it is hard to tell the accurate propagation paths and its characteristics. For example, sometimes we observe from the Earth that a CME is coming toward the Earth, but it just flies by instead of reaching the Earth. Therefore, it is possible to conduct imaging observation on ICME from the solar polar orbit, which is similar to typhoon observation by meteorological satellites. This kind of imaging observation and corresponding forecast of the ICME propagation direction becomes an important field of space weather observation.

(3) Observation and Modeling of Magnetosphere Dynamics

Since the launch of the first artificial satellite in 1957, we started to observe the magnetosphere and attempted to describe the configuration of the Earth's magnetosphere. In the 1980s, it was becoming clear that the magnetospheric configuration was not stable. Once a space weather event occurs, its configuration changes dramatically. For example, the altitude of magnetopause can be reduced from 10 to 6 Earth radii, and even geostationary satellites (36,000 km) can be fully exposed to the interstellar solar wind.

In the late 1990s, in order to understand the characteristics of the dynamic magnetosphere, relevant studies employing multiple satellites for multipoint detection of the magnetosphere were initiated. Since 2000, with the advancement of imaging technology, the idea of direct imaging of the dynamic magnetosphere has gradually become a reality. The Chinese geospace Double Star Program (DSP) together with ESA's Cluster Mission conducted joint six-point observation of the magnetosphere, as shown in Fig. 3.10. The Solar wind Magnetosphere Ionosphere Link Explorer (SMILE), an ESA-China Joint Scientific Mission, aims for the global imaging of the interaction between the solar wind and the Earth's magnetosphere to model the dynamics of the magnetosphere.

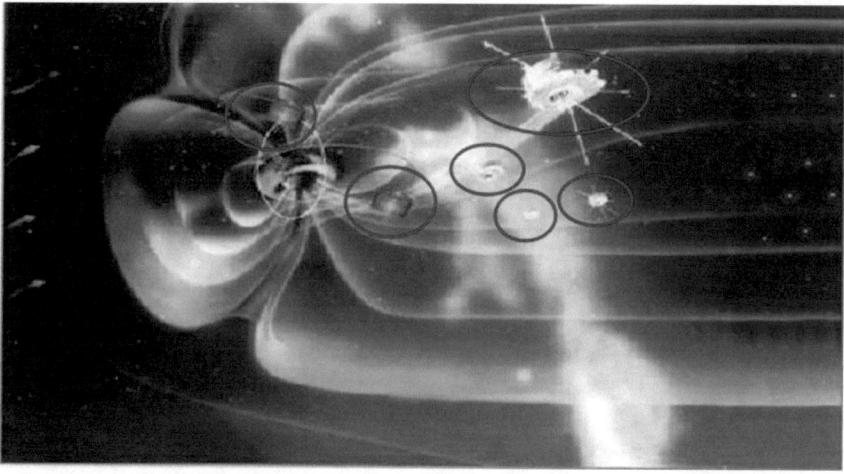

Fig. 3.10   China's Double Star Program and ESA's Cluster Mission realize joint six-point detection of the Earth magnetosphere, the first time in history

(4) Escape of the Earth's Oxygen Ions

Through the exploration of the Mars, scientists have found that there is continuous oxygen loss in the Mars atmosphere. If such a loss is reversible, and we rewind the clock back by a billion years, on the surface of Mars, there would have been waters several meters deep. This propels human being to think about the future of the Earth.

References

1. Tyson ND (2017) Astrophysics for people in a hurry. W. W. Norton & Company, New York
2. Committee on the Societal and Economic Impacts of Severe Space Weather Events, National Research Council (2011) Severe space weather events: understanding societal and economic impacts, a workshop report. The National Academies Press, Washington DC

Chapter 4
Major Frontiers in Space Science (II)

4.1 Introduction

In Chap. 3, the topics start with the universe and end with the solar system. In this chapter, the topics will shift from the solar system to the Earth. As mentioned in Chap. 1, compared with ground-based observations, observing the Earth from orbit gives a completely different perspective, which enables us to study the Earth as a planet, as well as a system. This is space Earth science. In this chapter, the Earth system will be introduced by elaborating on several spheres and cycles of the Earth, which will lead us to several important frontiers in space Earth science.

To study the kinetic properties of matter on the Earth, it is inevitable to encounter the issue of gravity, especially when matter moves upward or downward. For example, hot air being lighter than cold air must float upward, and the flame always goes up instead of going downward. In addition, the evolution of life on the Earth may somehow relate to the influence of the Earth's gravity. All these questions can be answered by in-orbit microgravity experiments, hence revealing the essential patterns among matter and the rules of life phenomena. After the introduction of the fundamental knowledge in these areas, the relevant major scientific frontiers will be introduced.

4.2 The Earth System and Its Future Changes

4.2.1 The Spheres and Cycles of the Earth

Taking the Earth as a global system, we can describe its systematic movements through its spheres and cycles. Traditionally, there are five spheres and circles, namely, the atmosphere, the ocean, the ice and snow, the lithosphere, and the biosphere. However, with the increasing space observation activities, now we have a more comprehensive understanding of the Earth and the concept of new spheres

© Science Press 2021
J. Wu, *Introduction to Space Science*, Springer Aerospace Technology,
https://doi.org/10.1007/978-981-16-5751-1_4

and cycles has been formed accordingly, which includes water cycle, energy cycle, carbon cycle, lithosphere, and biosphere. Of course, new spheres and cycles will also be re-categorized as we separate out the cycles of other physical elements, gaining new understanding of the Earth system in the future.

(1) Water Cycle

Water cycle, also called hydrosphere, as shown in Fig. 4.1, refers to the circulatory system of water on the Earth. There is a large amount of water in the atmosphere in gaseous, liquid, and solid state, which exchanges with the surface water through evaporation, rainfall, snow, and other forms. On the surface, there is water in liquid state, e.g., lakes and runoffs, and water in the subsurface. There is also water in solid state, e.g., snow, ice, and water ice in frozen soils. The surface water is replenished by precipitation and snowfall from the atmosphere. It also exchanges with the atmosphere, groundwater, and oceans through surface evaporation, sublimation, surface infiltration, and runoff. Below the surface, there are tens of meters, and even hundreds of meters of groundwater. Covering about 70% of the Earth's surface, the ocean reserves the largest amount of water on the Earth. The oceans transport a large amount of water vapor into the atmosphere through evaporation and is supplemented by the surface runoff and precipitation from the atmosphere.

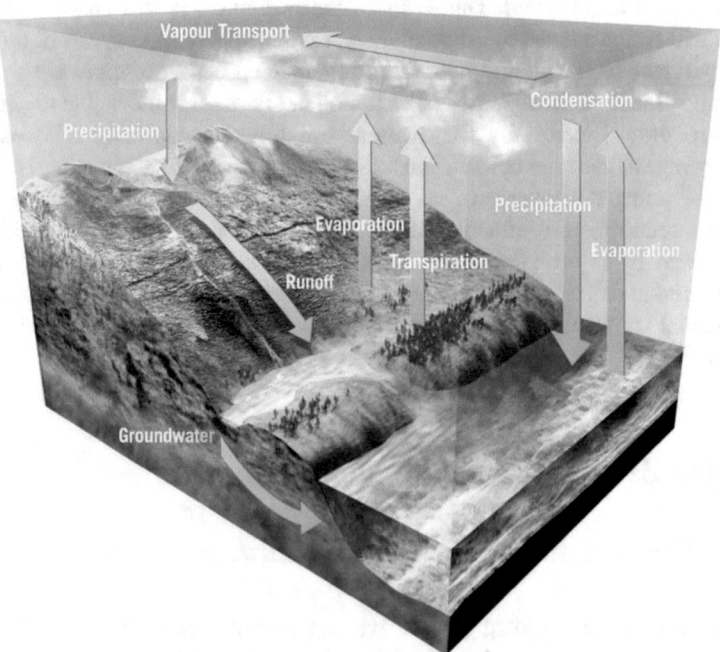

Fig. 4.1 Water cycle. Photo credit: ESA

Although the total amount of water in the atmosphere, the surface, and the oceans remains essentially constant in short terms, its distribution among the various processes is continually changing. The popular view of the science community [1] is that water movement should follow some patterns, such as the influence of the seasons, to form a circulation system. As a matter of fact, the circulation system is characterized by strong randomness and divergence, making it very unpredictable or it is difficult to use a mathematical model to describe the patterns.

To study the water cycle, it is necessary to observe the changes in the global water distribution from the vantage point of space. Using microwave and millimeter wave to observe the water cycle becomes the most effective method, which makes full use of the scale of water molecule and its characteristic of high conductivity. For example, the water contained in the soil surface (about 0.1 m in depth) can be directly inverted by using the L-band microwave, and the salinity of ocean water can also be inverted, making the study of the ocean current by tracking the changes of salinity possible. In addition, Ku- and K-band microwaves can be used to observe water droplets (rainfall) that have formed in the atmosphere and millimeter-wave observations can be used to invert water vapor and temperature in the atmosphere.

What's more, by measuring the changes of the Earth's gravitational field, we can indirectly invert the ground water information. The design of Gravity Recovery And Climate Experiment (GRACE) mission is based on this idea. Changes in the Earth's surface gravitational field as a result of the rise and fall of groundwater can affect the speed of a satellite. The GRACE spacecraft uses microwaves to measure the distance between two satellites flying one after another in the same orbital plane, so as to invert the changes in the Earth's gravity and corresponding rise and fall of groundwater. Based on the data obtained by GRACE, Fig. 4.2 shows the changes of the amount of water stored in Amazon Basin from January to December 2004.

(2) Energy Circle

The atmospheric cycle alone can be nominally described as the layers of atmosphere above the Earth's surface. But this description doesn't touch upon its functions, e.g. the constant water transfer. In addition, its own movement also propels the high and

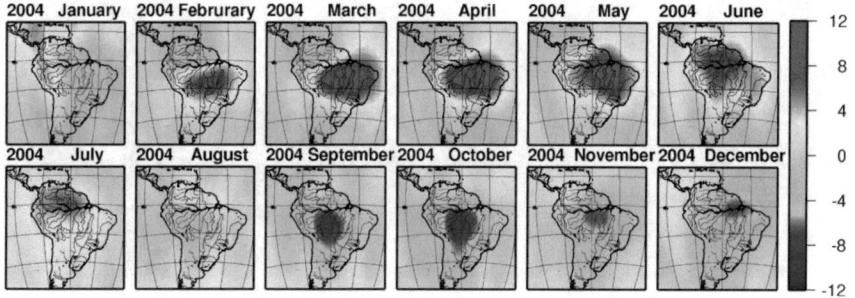

Fig. 4.2 Changes of the amount of water stored in the Amazon Basin varies from month to month in 2004 (unit: cm). Credit: NASA/JPL

low temperature movement across the globe. In fact, what we are more concerned about is the pattern of energy changes rather than the atmosphere itself. This is the energy cycle.

The energy cycle is mainly propelled by the temperature of matter and its motion in space, which is manifested as atmospheric temperature, surface temperature, sea water temperature, and the energy exchanges among them, as well as the patterns. Think of the Earth as a global system. If the Earth is constantly radiating outward and spreading the energy through the atmosphere into outer space without any energy or temperature inputs, its own energy would become less and less. But a general energy balance is stroked, because the solar radiation provides the Earth with an endless supply of energy. Figure 4.3 shows the Earth's energy budget. The temperature drops at night and rises again during the day due to the steady inflow of solar radiation, which peaks shortly after the noon. This happens every day. The input and output of energy, its distribution in different regions of the world, and its pattern of changes are the research questions for the energy cycle study.

As for the energy exchange brought by ocean temperature, the North Atlantic Current serves as a good example. Every year, the North Atlantic Current, characterized by warm temperature, brings warm current from the lower latitudes to the north part of the Atlantic Ocean, where it cools and goes downward before heading

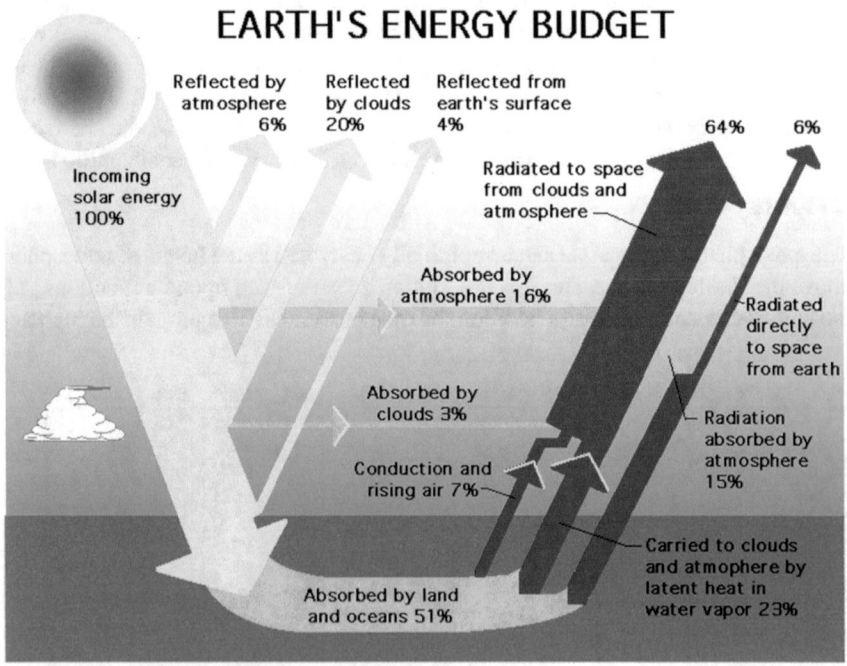

Fig. 4.3 The Earth's energy budget

to the lower latitudes, bringing heat to the higher latitude areas in Europe. In addition, there is also energy exchange between the troposphere and mesosphere. Some people believes that the temperature of the middle atmosphere is an indicator of the troposphere temperature. The temperature variations in the mesosphere are much greater than that in the troposphere. Experiments show that every 1 °C change in troposphere temperature will cause a 5 °C change in mesosphere temperature.

Since temperature is the most direct and measurable physical parameter of energy, remote sensing observation in infrared band becomes the main tool to study the energy cycle.

(3) Carbon Cycle

With the increasing attention on global warming, the research on carbon cycle gradually makes it a new research field in its own name, separating from atmosphere, energy cycle, and biosphere. The concentration of carbon dioxide in the Earth's atmosphere is incrementally increasing, blocking the surface radiation into space. The dwindling vegetation reduces the amount of carbon dioxide that plants can absorb from the atmosphere (known as a *carbon sink*). Human activities brought about burning of large amounts of plants and chemical fuels, resulting in an increase of carbon dioxide concentrations in the atmosphere. These factors are impacting the energy cycle, and their importance has already been recognized.

For the abovementioned reasons, the study of the Earth's carbon cycle has gradually evolved itself as an independent research field. As the carbon dioxide has absorption lines in the Near-Infrared Bands (NIR), using these absorption lines became the most effective means of observing carbon dioxide in the atmosphere. China has launched TanSat with the scientific objective to monitor with high precision the global carbon dioxide concentration distribution and its changes. Now the detection data are now available to the public. The global atmospheric carbon dioxide concentration as measured by TanSat is shown in Fig. 4.4. The observations in the visible spectrum can provide data to reveal the changes in surface vegetation coverage.

(4) The Lithosphere

Earthquakes and volcanic eruptions are major natural disasters, which are also the manifestation of the Earth's internal movement on the crust. The lithosphere studies large-scale movements of plates and predicts whether there will be intensive eruptions. In generally, the movements of plates are extremely slow, making the traditional measurements useless in observing the changes, let alone to tell the direction of its movements and its patterns. But since the birth of space technology, the study of the lithosphere has entered a completely new stage.

Using GPS to continuously monitor the position of continental plates for a long period of time to detect their movement is the observation method frequently chosen. For a certain region, the effective observation method is to use interferometric synthetic aperture radar to study its tiny movement and predict regional crustal movement such as landslides. The new trend driven by technology is to study changes in low-frequency electromagnetic fields as a result of the Earth's tectonic movements, and these changes are measured from space.

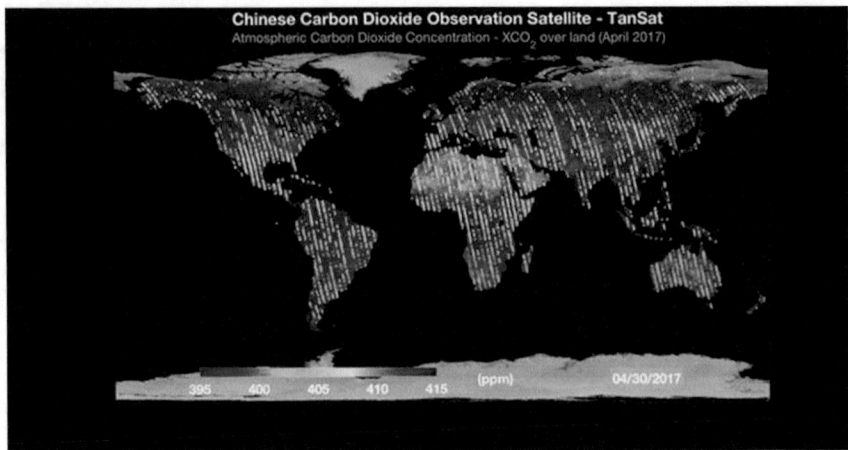

Fig. 4.4 Global carbon map (observation data of TanSat, April 2017)

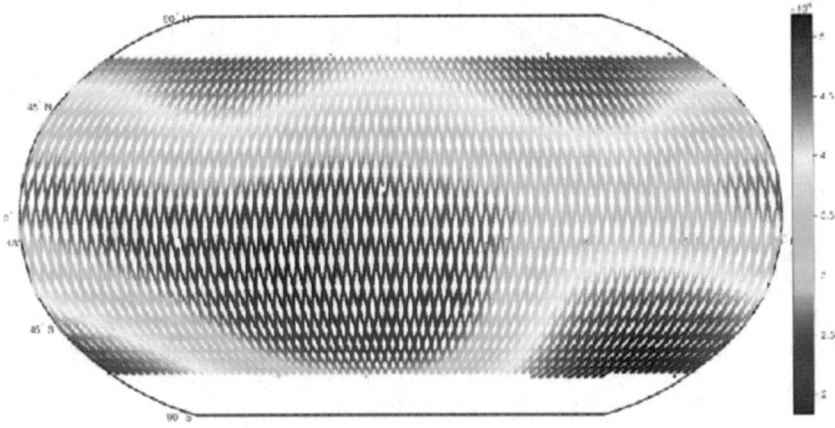

Fig. 4.5 Map the global geomagnetic (CSES)

Electromagnetic monitoring satellites aim to exclude the interference of space weather and study changes in the electric field caused by the pressing of tectonic plates. China Seismo-Electromagnetic Satellite (CSES for short, and also named *Zhang Heng-1*[1]) monitors the magnetic fields, electric fields, and particles. Its observations provide accurate data of the global geomagnetic field (as shown in Fig. 4.5) and the data will be open, step by step, to researchers worldwide.

---

[1] *Note* Zhang Heng (78–139) is a Chinese mathematician, astronomer, and geographer in the Han dynasty. He is best known for his seismoscope for registering earthquakes.

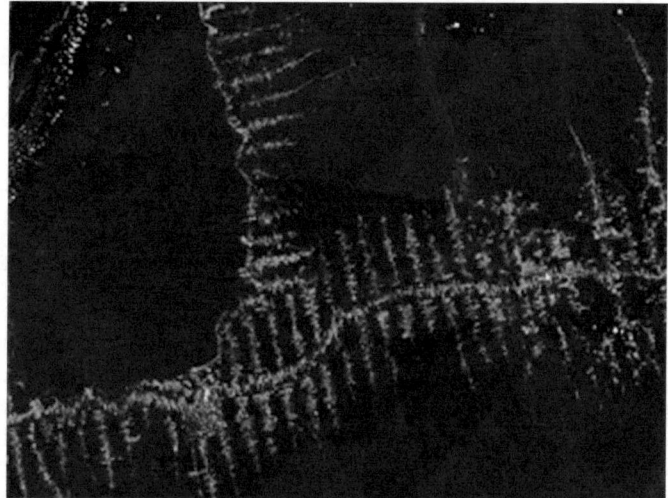

Fig. 4.6 First discovery of Amazon deforestation by NASA's satellite. Credit: NASA

(5) The Biosphere

The Earth's surface is covered with a great variety of plants and populated by diversified animals. They follow the rule of natural selection and the rule of survival of the fittest. They are also crucial indicators of global change. The impact of global change on the living environment of creatures is reflected in the survival and dying out of any species, and their migration and distribution, which is also the subject of the biosphere study. The migration of migratory birds, for instance, is a typical example of changes in animal distribution. In the past 100 years, the scope of human activities has been expanded enormously and the intensity also escalates, which has made profound impact on the distribution of the primitive biosphere. This, to a great extent, puts the future of mankind in danger. It is noteworthy that the decrease of Amazon Rainforest has drawn the attention from the public because of the satellite remote sensors. As shown in Fig. 4.6, satellite images reveal the fish-scale patterns in the Amazon Rainforest, which, as confirmed by later field investigations, turns out to be man-made paths to facilitate the transportation of timbers. In addition, satellite data is also used to study migration paths by placing GPS on migratory birds or other migratory animals.

4.2.2   Questions of Great Significance

(1) Climate Warming?

The well-known speech by Al Gore shows us the fact of global warming in the last 100 years. But, in fact, many regions have been experiencing temperature drop in recent years. This implies the uneven distribution of global change. Therefore, the questions are: will the global warming be a disaster for humanity?; is the global warming directly caused by human activities? For the long history of the Earth, in terms of temperature changes, there are many periods of high temperature, as well as periods of low temperature, or the ice age. During these periods, the impact of human activity is almost tantamount to nothing. So what are the causes behind these ice ages or high-temperature periods?

(2) Is it Possible to Continuously Observe the Changes of the Earth Over a Long Period of Time?

More than a decade ago, Group on Earth Observations (GEO) summarized more than 30 physical parameters that could be observed from space, and has been devoted to building a large Earth observation system by combining dozens of remote sensing satellites around the world, aiming to improve the comparability of remote sensing data. However, after more than 10 years' efforts, the difficulty of long-term, continuous Earth observation still lingers on. The main difficulty is data calibration. That is to say, which data should be reference data and how to unify the data are still to be determined. The difficulty lies in that the physical parameters for Earth observation from space are very complex, which is worsened by the fact that the performance of the instruments in space is incrementally declining, and the actual physical parameters from the ground are also changing not to mention the regional differences. The measurements conducted on the ground tend to bear no statistical significance, while space-based observations produce statistical parameters in larger scale. Therefore, how to calibrate the actual ground physical parameters with remote sensing parameters has always been a puzzling problem for the researchers. Other problems include the research time scale and lifetime of remote sensors. If a unified and long-term physical measurement standard cannot be established, it will be very difficult to study the various spheres and cycles of the Earth through continuous observation of several decades or even longer.

(3) Establishment of Mathematical Models for Each Sphere and Cycle

It is the dream of scientists to describe the water cycle, energy cycle, lithosphere, and biosphere through mathematical models. However, the Earth is a system so enormous and complex that no simple model can describe the patterns of its changes. A daunting job as it is, scientists are still working on it and many models have been built, which are now shifting from regional models to global models. For example, for the time being, we have clear understanding of the movement of the warm Atlantic current, as well as the patterns of El Nino and La Nina phenomena. If we continue in this direction, we must further strengthen our observations and constantly modify the models with the input of observation data. With the development of computer technology, more and more Earth system observation data has been entering into the database, which could be used as input for computers to build numerical models. These simulation models, instead of physical and mathematical models, may be a way out to effectively solve the problems plagued the Earth system.

(4) Is There Any Relationship Between Space Weather and Global Change?

As mentioned above, there have been many ice ages in the history of the Earth, when the impact of human activities was tantamount to nothing. Therefore, the question is what is the reason behind the cooling in much of the northern hemisphere? In recent years, with the strengthening of research on solar activities and space weather, awareness has grown in the relationship between space weather and the weather and climate in the atmosphere. As mentioned earlier, studies have shown that temperature changes in the troposphere will multiply in the mesosphere. For instance, an increase of 1 °C in the troposphere will lead to an increase of about 5 °C in the mesosphere. A change like this is sufficient to cause fluctuations in the mesosphere. At the same time, the mesosphere is also subject to the influence of solar activities, and changes will occur accordingly. Will these changes in turn cause temperature changes in the troposphere? The correlation between sunspot changes and the probability of flooding on the surface of the Earth had been studied in the history, but this is now considered ill-grounded. However, scientific research on the impact of solar activities on the atmosphere by establishing a chain of physical pathways from the surface of the Sun to the mesosphere has begun to receive widespread attention from the science community.

4.3 Microgravity Science and Space Life Sciences

4.3.1 How to Simulate Microgravity Environment

(1) Drop Tower

Drop tower (Fig. 4.7) is a building tens of meters or even hundreds of meters in height, with hollow space in the middle for free falling objects. The drop towers are usually built in the shape of cylinder. The air resistance will reduce the speed of free falling objects. Therefore, for some experiments that require higher microgravity level, it is necessary to install a vertical vacuum drop tube through which the experiment capsule will realize the free fall without air drag. Due to the height limitation, normally this kind of drop tower can only obtain a few seconds' microgravity state. There are more complex designs to extend the period of microgravity state, such as combining two types of drop facility: a drop well and a drop tower. In addition, an catapult system could be added, which accelerates the experiment capsule from the bottom of the tower, and after the acceleration stops, the capsule enters the microgravity state and continues to move upward for a period of time to reach the top of the drop tube before falling down, which can almost double the microgravity time. What needs particular attention is that due to the required high precision for the catapult system, it is necessary to ensure the verticality of the drop capsule inside the vacuum tube to make sure the capsule will not collide with the tube wall. In addition, the acceleration speed during the catapult is very high and there are possibilities that the overload may

Fig. 4.7 Drop tower of
Institute of Mechanics of
Chinese Academy of
Sciences

damage the test sample before the capsule is entering the microgravity state. Across the globe, the Center of Applied Space Technology and Microgravity (ZARM) in Bremen, Germany is the only drop tower that combines the catapult system and the drop system.

(2) Aircraft

Aircraft refers to the plane in which the pilot simulates a free fall parabola trajectory, as shown in Fig. 4.8. All the devices and apparatus inside the aircraft enter microgravity state during the falling process. The aircraft reaches its targeted lowest point, and then pulls up again, reaching its highest point, after which it swoops down again following a parabola trajectory, entering once again a weightless state. Such practice can be done a couple of times in a single flight. When the plane is pulling up, it enters super gravity state, which, however, may also impose a certain degree of disturbance to the experimental samples. Since the maximum altitude that the aircraft can reach is only around 10,000 m, the period of microgravity state achieved is only about 30 s in a parabolic flight. This also requires the pilot should have a high level of driving skills.

(3) Balloons and Sounding Rockets

Balloons (e.g., high-altitude balloon, Fig. 4.9) and rockets (e.g., sounding rockets, Fig. 4.10) for microgravity experiments follow the same procedure that the experiment packages are taken to an altitude from tens of kilometers to hundreds of kilometers and then are dropped for free fall. But the experiment effectively stops when the experiment package falls to an altitude of 30 km where the atmospheric density gradually increases and the air resistance generated destroys the microgravity environment. Taking into account that the maximum height that the high-altitude balloon can fly is just around 40–50 km, the period of microgravity obtained in this way does not increase significantly. In comparison, a few minutes to ten minutes of high-quality microgravity time can be obtained by using sounding rocket for microgravity

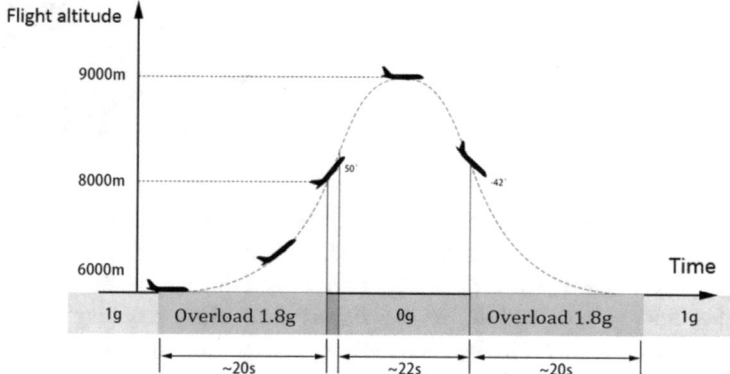

Fig. 4.8 Parabola aircraft

Fig. 4.9 High-altitude balloon

Fig. 4.10 Sounding rockets

experiments. The sounding rockets, however, cost more. Both ESA and NASA use sounding rockets to conduct microgravity and life science experiments, but the experiments conducted each year are ranging from several times to no more than a few dozen. For microgravity and life science experiments using sounding rockets, the samples are to be recovered. Therefore, the sample recovery technology should be considered at the same time. Most sounding rocket launch sites are located in the less populated areas and parachutes are used for the soft landing of samples. The landed samples contain accurate positioning information which can facilitate the speedy sample recovery by helicopters or cross-country vehicles. Very few sounding rocket sites, like the one in Norway, use offshore recovery methods.

(4) Recoverable Satellite

The man-made satellite will stay in orbit for continuous operation without crashing into the Earth atmosphere on the condition that the centrifugal force (an apparent outward force) of the orbiting satellite equals to the gravitational pull of the Earth. Under the conditions of the Earth's gravitational field, such an orbital velocity is called the first cosmic velocity. Depending on the altitude of the orbit, the speed varies from 7.6 to 7.8 km/s. Once a satellite reaches this orbital velocity, it acquires a simulated microgravity environment as the centrifugal force offsets the Earth's gravitational force. Although the microgravity environment is simulated, its existence is as long as the orbital lifetime of the satellite, which could be measured by weeks, months, and years. Therefore, artificial satellites are the best experiment platform for long-term microgravity and life sciences experiments [2]. It is noteworthy that the sample recovery will bring additional technical challenges, such as the protection of the re-entry cabin (recovery module) from aerodynamic heat after entering the atmosphere, the parachute opening before landing, the uncertainty of the landing zone after parachute opening, and the positioning and swift recovery of samples after landing. Compared with sounding rockets and artificial satellites that do not require recovery, the cost of recoverable microgravity experimental satellites is higher.

(5) Manned Space Laboratory and Space Station

Compared with recoverable microgravity satellites, the microgravity and life sciences experiments in manned missions involve human factors, which can significantly improve the efficiency of experiments and can better cope with randomly happened experimental phenomena. Therefore, more advanced and more sophisticated experiments can be conducted at a greatly increased cost. Therefore, for manned space laboratories and space stations, the study and selection of proper experiments with scientific objectives of great impact remains a critical challenge for microgravity and life science teams. Figure 4.11 is an artistic concept of China's future space station.

4.3.2 What Changes Under Microgravity?

Liquids performance in weightless environment is the most eye-catching. The bubbles in the water will move in all directions instead of moving upward. For mixed liquids, the distinguishable liquid layers due to their different specific gravity will no longer exist. A drop of water will float in weightless environment, and the surface tension replaces the gravity to become its most important force.

In microgravity environment, the hot air is no longer moving upward, so the flame doesn't go straight up and its shape will be circular in still air.

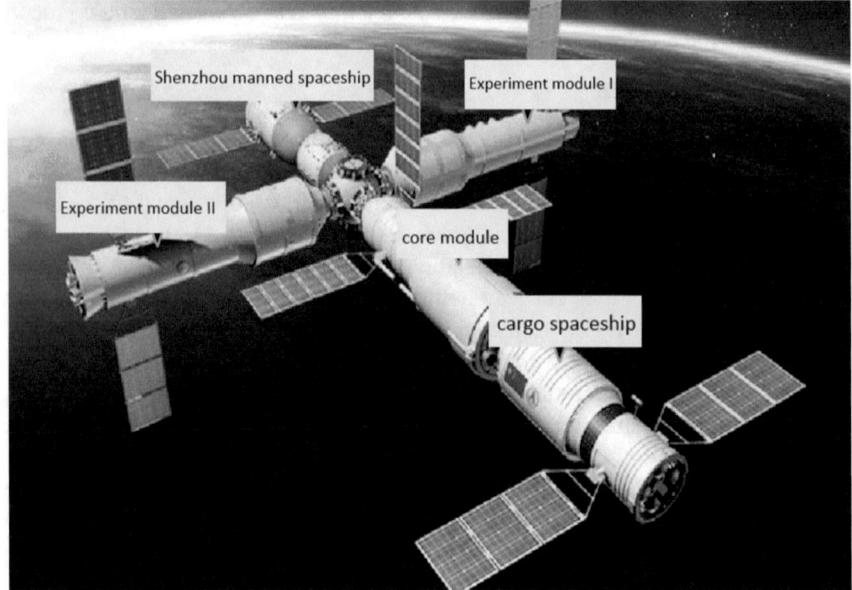

Fig. 4.11   An artistic view of China's future space station

In zero gravity, plants will lose the momentum to grow upward, and since the water in the soil will not move downward, the same is true for the roots which will grow in all directions.

The cell, in nature, is liquid living organism enclosed by the cell wall. Therefore, it follows the rules of the liquid in the microgravity environment. Advanced lifeforms will soon suffer bone loss. Necessary in-orbit exercise for astronauts will, to some extent, slow the process of bone loss.

In short, all physical phenomena associated with gravity will disappear. Surface tension and interface adhesion will become the dominant forces.

4.3.3   Biological Radiation Effect

For living things, the biological radiation effect is another special effect when entering space. When irradiated by space particles, the Deoxyribonucleic Acid (DNA) will break, causing genetic variation. However, caution that the well-known super big space cucumber and tomato are not the necessary product of all the seeds after space flight, but the product of the outstanding seeds that has undergone screening and cultivation for several generations.

In view that radiation can change the genes of seeds, can seeds be irradiated at ground accelerators for future planting? Experiments have found that the effect of ground accelerators on seeds is not the same as that in space, which cannot replace the role spaceflight played on the seeds. However, the basic theories and principles of these phenomena remain elusive.

4.3.4 Fundamental Physics Experiment

In space, it is also possible to carry out some physical experiments that cannot be carried out on the ground or require higher precision, such as the verification of the equivalence principle, Bose-Einstein Condensation (BEC) experiments, high-precision cold atomic clocks, long-range distribution experiments of quantum entanglement, etc. These experiments are complicated and some are not necessarily related to microgravity. For example, China's QUantum Experiments at Space Scale (QUESS), nicknamed Micius, has carried out large-scale high-speed quantum key distribution experiments, quantum entanglement distribution, and quantum teleportation experiments. Although not related to microgravity, these experiments fall in the scope of fundamental physics experiments.

4.3.5 Questions of Great Significance

(1) Convection, Diffusion, and Transport of Complex Fluids

Under the condition of microgravity, the fluid will change its pattern of motion. When the composition and boundary conditions are complicated, our research focuses on the patterns of motion of fluid under microgravity, such as the stratification, diffusion, and transport of multiphase liquids. These motion patterns bear significant application values in many areas, such as residual fuel in rocket engine fuel tanks, liquid flow in spacecraft heat pipes, the movement of lunar dust particles at 1/6 g after landing on the lunar surface, and the movement of liquid in cells of living things.

(2) Genetic Variation of DNA After Particle Radiation

As mentioned above, when exposed to particle radiation, an arm or section of the DNA double helix structure can break. The important answers to a couple of questions remain elusive, such as in which section and segment the break will occur, whether it will repair itself, and whether it will inherit and affect future generations?

From a macroscopic aspect, there are additional issues related to particle radiation, such as astronaut health and spaceflight breeding.

(3) Does Darwin's Theory of Evolution Still Work Beyond the Planet Earth?

It is self-evident that mathematics is universal in the cosmos. That one plus one equals two holds true in any other potential extraterrestrial civilization despite the possible different expressions. For the time being, it seems that the basic laws of physics that we have summarized so far (not including those related to gravity) are also universal. For example, Einstein's general theory of relativity holds true to explain all the galaxies we have observed, no matter how far away they are.

But the basic laws associated with life sciences, such as Darwinian evolution, have not been rectified beyond the Earth. This is largely because we haven't found any signs of life in our solar system, not to mention the worlds outside the solar system. Therefore, sending Earth life to space beyond the Earth, such as the space station or the moon, to carry out long-term research on the basic laws of life science becomes the frontier of space life sciences.

(4) How to Simulate Microgravity on the Ground?

As mentioned above, drop towers, aircraft, and balloons and sounding rockets represent some ways to simulate the microgravity environments. However, none of these ways can obtain longer periods of microgravity. A large number of experiments have to be done in space at much higher cost. Even for experiments on a manned space station, there are many interference and disturbance we need to tackle, such as the vibration of various instruments and fans inside the station. Even the rotation of flywheels on a satellite can reduce the microgravity level. Currently, the microgravity levels for microgravity satellites are from 10^{-3} to 10^{-5} g. For microgravity satellite missions, the smooth sample return should be guaranteed, which increases the experiment cost.

At present, two immature ground simulation methods are being developed, namely, magnetic suspension and Rotary Cell Culture System (RCCS). However, neither method can fully simulate the real microgravity environment. Therefore, it is necessary to carefully analyze the objectives and requirements of the experiment before choosing the method to simulate the microgravity environment.

References

1. Lutgens F, Tarbuck E (2016) Foundations of Earth science, 8th edn. Pearson, London
2. Wenrui H (2010) Introduction to microgravity science. Science Press, Beijing

Chapter 5
Space Systems Engineering and Its Systems

5.1 Introduction

As a new interdisciplinary subject, space science relies on spacecraft to conduct related research. Therefore, it is necessary for researchers in space science to understand the basic ways to gain access to space and have the rudimental knowledge of space systems engineering.

Space engineering is systems engineering in nature, taking in view of its complexity, high risk, and high cost. In addition, since the beginning of space age, space engineering is highly related to politics and bears great social benefits. Space science satellite engineering belongs to space systems engineering and shares the afore-mentioned characteristics. In this chapter, the general characteristics of space systems engineering, as well as the components and functions of each system, will be briefly introduced.

5.2 Space Systems Engineering

In general, a space engineering mission consists of five major systems: spacecraft (satellite), launcher (rocket), spacecraft launch facility (launch site), Tracking, Telemetry and Control (TT&C) system, and user application (ground application). For particular space engineering missions, such as manned space engineering projects, there are additional systems, e.g., the astronaut system and recovery site system. For missions that require several coordinated spacecraft, such as lunar and Mars landing missions, the spacecraft system will be divided into sub-systems, e.g., orbiter, lander, and rover. As a result, the abovementioned five systems are the minimal and core systems that could be added according to the nature of different missions.

Space systems engineering is characterized by its large scale, sophisticated technology, high reliability requirements, high cost, long development period, and

© Science Press 2021
J. Wu, *Introduction to Space Science*, Springer Aerospace Technology,
https://doi.org/10.1007/978-981-16-5751-1_5

prominent social and economic benefits. Some typical space systems engineering missions include China's first artificial satellite Dongfanghong-1 (DFH-1), China's manned space program, China's lunar exploration mission Chang'e-1, etc. The Apollo program that realized America's manned lunar landing and America's space shuttle program are typical examples of modern space systems engineering. The characteristics of space systems engineering are as follows.

5.2.1 Complexity

First of all, the complexity of space system engineering is manifested distinctly in the large number of employees in the space sector. A typical space program can involve as many as thousands or even tens of thousands of people. A more astounding example is the US Apollo program which involves, directly and indirectly, hundreds of thousands of participants. Apparently, with the technology advancement and the extensive use of computers, as well as the commercial mass production of a large number of components, an engineering program for a mature application satellite, at present, involves only a few hundred direct participants. Even that the complexity of coordination and collaboration between systems as required by the systems engineering still exists.

Secondly, the complexity of space system engineering is also manifested in its wide geological coverage. Experiments related to the spacecraft must be carried out at large ground-based experimental facilities and before entering the satellite launch site, there will be long-distance highway, railway, or air transportation. In terms of telemetry, tracking, and data receiving, the ground stations need to be deployed across the nation and even over the globe. What's more, in many cases, the tracking ships are required to be positioned in designated areas. These cross-region arrangements and coordination also reflect the complexity of space system engineering.

Furthermore, the complexity of space systems engineering is reflected in its long life span. In general, a satellite mission takes 10 years from mission proposal, approval, mission design review, testing, production of engineering qualification, and flight model production, all the way to launch and operations. For large projects, the timeframe is even longer. The Double Star Program (DSP), China's first space science mission, takes 12 years from mission proposal in 1997 to its end of operation in 2009. For the first phase of China's manned space program, it takes 11 years from the approval by the central government in 1992 to Yang Liwei's entry into space in 2003. There are missions that take even longer time, such as the Cassini-Huygens mission, a collaboration between NASA, ESA, and the Italian Space Agency (ASI), which takes 32 years from the mission study in 1985 to the end of operation in 2017 due to the exhaustion of propellant.

Finally, the complexity of space programs is also reflected in the complexity of its various systems. A space engineering mission usually consists of the following five major systems: spacecraft (satellite), launch vehicle, launch facility, TT&C system, and user application. Each system is further divided into a hierarchy of subsystems.

In terms of the disciplines involved in space systems engineering, it shows prominent interdisciplinary nature. The disciplines and majors involved include but not limited to orbit dynamics (mathematics, mechanics), attitude dynamics (information science, mechanics), propellant chemistry (chemistry, combustion science), mechanical structure mechanics (finite element computational mathematics, mechanics), electronics (theory of electromagnetic field and microwave technology, electronics, computer science, microelectronics), and the knowledge of space physics, space environment, particle physics, etc. In recent years, due to the extensive use of computer chips, the software on-board the spacecraft has become increasingly complex, and software engineering has been applied intensively into various space engineering systems and its subsystems.

5.2.2 High Risk

Of all the complex and sophisticated systems engineering, space systems engineering is the most risky. Once the launch vehicle is ignited and lifts off, it is impossible to stop it. In the early days of human space exploration, many accidents occurred, e.g., the explosion upon the firing of rockets engine, explosion after the liftoff, crashing into the Earth atmosphere upon failing to reach the orbit, and deviation from the target orbit. Failures are declared for those missions that the malfunctions occurred on the in-orbit spacecraft which is impossible to be repaired, leading to its failure to meet the designated mission objectives. During the in-orbit mission operations, there is possibility that the electronic components of the spacecraft may be damaged by the radiation of high-energy particles in space, causing major malfunctions or even failures of the missions. According to incomplete statistics, in the last 20 years, the number of failed spacecraft due to launch accidents and other malfunctions still accounts for 5–10% of all launched missions. The biggest accident in the history of human spaceflight occurred at the Baikonur cosmodrome[1] in the Soviet Union, when a rocket exploded at the launch pad, killing hundreds of people including generals and soldiers. The people who know what happened at the launch pad are all killed, leaving finding the exact reason of the explosion impossible mission and up until now the exact reason still remains elusive. Of the 135 space shuttle missions, there are two major accidents on the Challenger (Fig. 5.1) and Columbia space shuttle. The Challenger space shuttle exploded shortly after liftoff, while Columbia space shuttle crashed upon return to the Earth. The two failed missions killed a total of 14 astronauts. Malfunctions, big or small, are very likely to happen even after the satellites enter into orbit, accounting for more than 60% of all missions. An American Mars lander, which messed up the meters and miles in the system design, ended up crashing on the Mars upon descending due to severe navigation errors.

---

[1] Baikonur Cosmodrome located in Kazakhstan is currently operated by Roscosmos, the Russian Space Agency.

Fig. 5.1 On January 28, 1986, the challenger space shuttle exploded shortly after liftoff, killing all seven astronauts on-board. Photo credit: NASA

Therefore, the management measure in the implementation of space system engineering is crucial for the identification of risks in all the phases of design, test, and production, and it plays crucial role in formulating effective measures to tackle and reduce risks.

5.2.3 High Cost

It usually costs tens to hundreds of millions of Chinese yuan (RMB) to produce a launch vehicle, and the normal cost of developing a satellite is hundreds of millions to billions of Chinese yuan. There are additional cost for launching a satellite into orbit and the follow-up continuous operations, such as the costs of launching, telecommunication and telemetry, and ground operations, which normally amounts to tens to hundreds of millions of Chinese yuan, even if we exclude the cost for technical improvement at regular intervals and the necessary construction costs. Therefore, the total estimated cost for a conventional space engineering mission is around 1 billion Chinese yuan or even more, while larger missions cost billions of Chines yuan. Manned space missions cost even more. Taking China's manned space program as an example, it costs about 20 billion Chinese yuan since the project's approval to the launch of the first astronaut into space. The James Webb Space Telescope (JWST) to be launched in 2021 is NASA's largest space science mission after the Hubble Space Telescope (HST), and it is expected to cost nearly 10 billion US dollars. It is justified to say that space systems engineering missions are extremely costly, which explains that it is usually implemented by the government.

In recent years, due to the technological advancement, small satellites and microsatellites programs using relatively cheap commercial off-the-shelf components

(COTS) began to emerge, and these missions are usually piggybacked. The emergence of CubeSat using standard components, in particular, makes it possible that college students can directly participate in and implement low-cost satellite missions with short development cycles and fast pace. For the time being, the cost of a CubeSat mission approved by the universities in the United States can be as low as tens of thousands of dollars excluding the labor cost which usually involves only teachers and students. However, due to its limited functions, it is difficult to implement a mission that bears substantial practical significance. What's more, as a student experiment mission, 1 million yuan is also costly.

Therefore, space systems engineering remains, on the whole, a high-cost activity with national agencies as the dominant players.

5.2.4 Sensitiveness to Political and Social Benefits

In the early days of space age, only two superpowers had the capability to launch satellites and explore space. Therefore, space evolves into a prominent arena where they rival with each other and demonstrate their national strength. From the late 1960s to the 1970s, France, Japan, China, and UK joined in the club one by one. At present, more than a dozen countries in the world have the capability to develop and launch satellites, while dozens of countries have the capability to develop satellites, and more countries have participated in, to varying degrees, space missions led by other countries. But launching a spacecraft into orbit is still a symbol of national economic, scientific, and technological strength, which has very obvious political and social benefits. In retrospect, President Kennedy announced in person the implementation of the Apollo program; President George W. Bush and President Trump have also signed the U.S. space policies or papers into effect, announcing the United States' return to the moon and eventually sending humans to Mars. In China, every time before launching astronauts into space, the top officials of the nation would meet with them to see off. From the abovementioned points, we can see that the governments attach great importance to space systems engineering. Recently, the United States has established the Space Force to ensure security in space defense. A considerable part of China's space budget also goes to national defense. In terms of scientific exploration, more than 10 American scientists have won the Nobel Prize in physics by mining the data of scientific satellites, which indicates that space science missions can also have a certain social impact.

Due to the political and public impact of space systems engineering, most of the space programs in the world are funded by the government. The government's fund comes from the taxpayers. Therefore, it is necessary to report the implementation of the mission to the public to explain to taxpayers how the fund is used. For the missions that are scientific and exploratory in nature, the effort for public outreach to spread knowledge to the public is made during the mission implementation. This again enhances its social benefits and increases public participation, making it more socially sensitive. For manned space program, it involves human activities in space, and every move of astronauts in space is under the public's attention, which increases their sensitiveness to social benefits.

In recent years, commercialization has been fully realized in some space applications, such as communications satellites, television broadcasting satellites, and some remote sensing satellites. Due to their wide service coverage and the fact that their users are scattered all over the world, the operation status of satellites has close relations with the public, and any malfunction of the satellites will cause concern or even public panic. Taking an American malfunctioning communication satellite, for example, the malfunction leads to the situation that many banks around the world lost contact with their users, and the users failed to receive expected information, delivering certain blow to the society and the economy. In addition, if there is no satellite to provide weather forecast, it will be difficult to forecast the typhoons and make necessary warning on its movement, which will also bring considerable economic losses to the society. Some space engineering missions are international cooperation projects in nature. Such cooperation must be based on political mutual trust among the countries. International cooperation in space cannot be materialized without the support of the participating governments, which also makes these cooperative missions highly politically sensitive.

5.3 System Components of Space Systems Engineering

In general, a space engineering mission usually consists of the following five major systems: satellite/spacecraft, launch vehicle, launch facility, TT&C system, and user application as shown in Fig. 5.2. These systems are not rigid and are subject to changes according to mission requirements and work division. For example, in the manned space system, the astronaut system, recovery field system, etc. can be added. This section focuses on the aforementioned five systems which are the most conventional and the most basic.

Fig. 5.2 The five major systems of space systems engineering

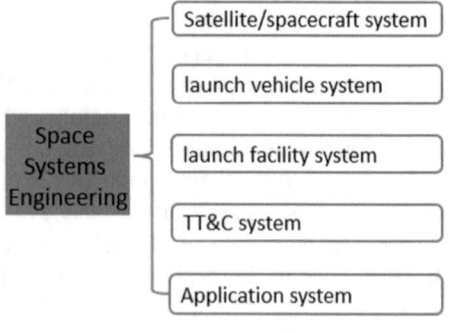

5.3.1 Satellite/Spacecraft System

Spacecraft collectively refer to satellites, manned spaceship, space stations, solar system probes, landers, rovers, etc. This section mainly introduces the satellite system.

In space systems engineering, the satellite/spacecraft system plays the core role and is the most important part for the whole mission [1]. For application satellites, the satellite/spacecraft system refers to the platform carrying the application payloads. The satellite/spacecraft system is usually divided into minimum eight subsystems according to mission requirements, including System Team, Structure and Mechanism, Power, Attitude and Control, Thermal control, Central Computer, Communication, and Harness. These subsystems constitute the main body of the platform and they are indispensable, regardless of mission types. In addition to these subsystems, the payload system is another important subsystem for the mission. For communication satellites, the payloads are communication antennas, transmitters and receivers, transponders and processors, etc., while for remote sensing satellite, the payloads are optical camera, microwave remote sensor, etc. For science satellites, the payloads, depending on the mission objectives, may be telescopes, detectors, experimental instruments, and so on.

A subsystem that is closely related to the spacecraft launched into orbit is the ground test system, where the facility simulates the tracking and telemetry by the ground stations to connect the communication facilities and the spacecraft. This subsystem participates in the whole satellite development process.

In addition, the spacecraft in development are required to undergo various simulated space environment tests, which are carried out by the environment simulation laboratories on the ground. Although they are not part of the satellite system, they play an important role in its development.

5.3.2 The Launch Vehicle System

As the most important foundation of space systems engineering, the launch vehicle system is the core technical facility to overcome the gravity of Earth. In the early days of space age, the most important technological breakthrough is the development of a launch vehicle capable of carrying payloads with a certain mass and of accelerating satellites to the first cosmic velocity.

Engine is the core of the launch vehicle. In this line of reasoning, the engine system is the most important subsystem for launch vehicles, which is followed by the subsystems of Structure, Control, Propellant, and Ground Test and Launch Control, etc. In order to reduce costs, ESA once proposed to reduce the number of rocket engines. The ariane-5 rocket's core stage only has one high-thrust engine.

For space engineering missions, the launch vehicle system is the most risky. When it comes to manned space program, once the launcher explodes, it will seriously threaten the lives of astronauts and even the ground staff is also in peril. Even the failure of robotic space mission will cause huge loss to individual country.

Since the successful launch of China's first man-made satellite on April 24, 1970, China's Long March (CZ) series has developed into a family series with diversified launch capacities. Before the CZ-5, China's rocket engine technology is literally based on that of the Soviet Union, which inherited the technical foundation of the V2 rocket engine developed by von Braun's team during the World War II. The engine propellant is toxic Unsymmetrical Dimethyl Hydrazine (UDMH). In case of propellant leak or a major accident at the launch site, the ensuing pollution will threaten the lives of the staff. As a kind of green and environment-friendly heavy rocket, CZ-5 uses non-toxic liquid hydrogen and oxygen as propellant, a new technology that is widely adopted by major space faring countries in the world. In terms of launch capacity, the CZ-5 has reached the capacity of 25t in low Earth orbit and 14t in geostationary transfer orbit, which matches the capacity of mainstream rockets in the world. The United States and Russia are also stepping up the development and experiment for the new generation super heavy launcher that can reach low Earth orbit with launch capacity of dozens of tons.

The Chinese manned space program has more than a dozen launches so far, with only one blunder happened in the early days. The unmanned Shenzhou 2 reentry capsule's parachute failed to open upon landing. According to the China Academy of Launch Vehicle Technology (CALT), by the end of 2018, China's CZ launch vehicles had made 194 flights, with only 11 failures (including 10 total failures and 1 with partial success), producing a success rate of more than 94%. Among the launch vehicles in the world, the CZ series have a relatively high success rate.

See Table 5.1 for detailed parameters of some CZ series.

5.3.3 The Launch Site System

For spacecraft, the launch site is their gateway to space. Facilities at the launch site are of crucial importance to launch vehicles using liquid propellant, i.e., they can provide necessary support and service for the rocket assembly, testing, propellant storage, refueling, weather forecast, and tracking and telemetry before entering the intended orbit. During the launch campaign, there are countdown procedures, such as 3 h preparation, 90 min preparation, 60 min preparation, 30 min preparation, 5 min preparation, 1 min preparation, and the last 10 s countdown.

Jiuquan Satellite Launch Center located in Gansu Province is China's first launch site. It is the launch site for China's first artificial satellite and the current launch site of China's manned space program. It has the capacity to support astronaut training and various preparations before launch, as well as a rapid escape from the launch tower in the event of a contingency.

Xichang Satellite Launch Center, located in Sichuan Province (Fig. 5.3), is China's first launch site dedicated to the launch of Geostationary Earth Orbit (GEO) satellites. Since geosynchronous satellites are always positioned above the equator, launching

Table 5.1 Parameters of some Long March (CZ) launch vehicles

Designators		Propellant type	First successful flight	Length (m)	Diameter (m)	Stages	Liftoff mass (t)	Liftoff thrust	Launch capacity (t)	Remarks
CZ-2	C	Liquid	1982.09.09	43	3.35	2	242.5	2961.6KN	4.1 (LEO)	The launch vehicle for the FSW recoverable satellite program (science and technology experiment, seeds satellites), Haiyang-1 (Ocean series), CFOSat (Chinese French Oceanic satellite), Huanjing-1 (environment)
	F	Liquid	1999.11.20	58/52	3.35	2	498/493	5923.2KN	8.1/8.6 (LEO)	• Has successfully launched 11 Shenzhou spaceships, 1 Tiangong-1 space lab and 1 Tiangong-2 space lab • The total length of the rocket is 58 m for manned mission and 52 m for target spacecraft; liftoff mass 498t for manned mission and 493t for target spacecraft; launch capacity 8.1t for manned mission and 8.6t for target spacecraft

(continued)

Table 5.1 (continued)

Designators		Propellant type	First successful flight	Length (m)	Diameter (m)	Stages	Liftoff mass (t)	Liftoff thrust	Launch capacity (t)	Remarks
CZ-3	A	Liquid	1994.02.08	53	3.35	3	243	2961.6KN	2.6 (GTO)	Has launched Dongfanghong-3 (DFH-3), Zhongxing-23 (Sinosat-23), Fengyun-2 series, Chang'e-1 and Beidou-2 navigation satellites series, etc.
	B	Liquid	1996.02.15	56	3.35	3	456	5923.2KN	5.5 (GTO)	Has launched communication and remote sensing satellites from home and abroad, solar system probes, BeiDou navigation satellites, etc.
	C	Liquid	2008.04.25	56	3.35	3	368	4442.4KN	3.9 (GTO)	Has launched communication and remote sensing satellites, solar system probes, BeiDou navigation satellites, etc.
CZ-5	/	Liquid	2016.11.03	57	5	2	869	10524KN	14 (GTO)	The launcher for lunar exploration program, mars exploration program, etc.
	B	liquid	2020.05.05	54	5	1	837.5	10524KN	25 (LEO)	The launcher for manned space station mission

(continued)

Table 5.1 (continued)

Designators		Propellant type	First successful flight	Length (m)	Diameter (m)	Stages	Liftoff mass (t)	Liftoff thrust	Launch capacity (t)	Remarks
CZ-7	/	Liquid	2016.06.25	53.07	3.35	2	594	7200KN	14 (LEO)	The launcher for cargo spaceship mission
	B	Liquid	2020.03.16	60.1	3.35	3	573	7200KN	7 (GTO)	As the main new generation launch vehicle for high-orbit missions, it will be mainly used for GTO, LTO missions, etc.
CZ-8		Liquid	/	53.34	3.35	2	359	4800KN	8.1 (LEO)	• As the main new generation launch vehicle for medium and low orbit missions, it will be the launcher for defense, civil, and commercial missions, e.g., National Satellite Internet System and medium and low orbit remote sensing satellites, etc. • Intended first flight in 2020[5.1]
CZ-11		Solid	2015.09.25	21	2	4	58	1200KN	0.7 (LEO)	As a new generation of launch vehicle for rapid response launches, it will be the launcher mainly for small size defense, civil and commercial missions in low and medium orbits

Note Data from *CALT Launch Vehicle Manual-2019 Edition*
5.1 *Note* CZ-8's maiden flight was successful on December 16, 2020

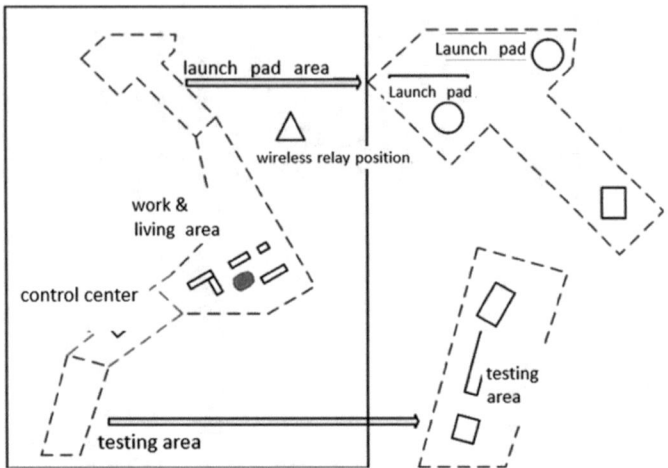

Fig. 5.3 Layout of Xichang Satellite Launch Center

from lower geographic latitude has the advantage of reducing orbit maneuvering, hence saving propellant. The geographical latitude of Xichang Satellite Launch Center is 28.5°. It is also the country's first low-latitude launch site.

In addition, China also has the Taiyuan Satellite Launch Center in the western part of Shanxi Province, and the Wenchang Satellite Launch Center in Hainan Province which has been put into operation in recent years. Wenchang Satellite Launch Center is China's first launch site for non-toxic and pollution-free launch vehicles, such as the CZ-5, CZ-6, CZ-7, etc. Its geographic latitude is 19°, closer to the equator, so it is an ideal place to launch low-inclination Earth satellites (such as geostationary satellites) and solar system probes. What's more, China's future space station will be launched from Wenchang, while the Shenzhou spaceship will still be launched from the Jiuquan Satellite Launch Center. The reason is that the space station's diameter is too big for the railway transport, and only Wenchang can be accessed by ship. Launching from seaside launch site is also the normal international practice, because when choosing the location of a launch site, consideration should be made on the landing area of first-stage rocket remains to make sure the minimum harm to local residents. Therefore, the landing zone for Wenchang is safer, which is also an advantage. Wenchang Satellite Launch Center, as China's first launch site open to the public, is also more accessible geographically.

The National Space Science Center of the Chinese Academy of Sciences has a sounding rocket launch site in Danzhou, Hainan. It's the only launch site in China dedicated to sounding rocket launching, which is also a comprehensive, low-latitude space environment exploration base.

5.3.4 TT&C System

Whether the launched spacecraft can reach the first cosmic velocity at the targeted position, whether it can successfully separate from the launch vehicle, and whether it can obtain the correct attitude upon separation all rely on the support of the Tracking, Telemetry and Control System (TT&C). As for China's first man-made satellite, the requirement for the TT&C system is "visible, audible, and trackable". For the launch vehicle, its success is determined by whether the spacecraft can enter the designated orbit. The report of the TT&C system indicates whether the launch vehicle system is successful. After the spacecraft entering the orbit, TT&C system will stay in tune to continuously track it. The responsibility of the TT&C system includes inputting uplink mission instructions, calibrating time baseline, receiving downlinked spacecraft engineering parameters (housekeeping data), etc.

The TT&C system normally comprises the following subsystems, including tracking and telemetry, unified time, trajectory calculation, display and recording, communication and data transmission, etc. The tracking and telemetry system communicates with the spacecraft via large ground antenna, and the communication is realized through the designated wireless (microwave) frequency. It is also responsible for sending the uplink instructions and receiving downlink engineering data. The unified time system is responsible for synchronizing the time within the system and upload the time to spacecraft. The trajectory calculation system is responsible for the calculation of trajectory parameters, pass time, etc. The display and recording system is responsible for displaying all parameters in real time to the staff on site as well to the users. It is also responsible for recording and storing relevant data. The communication and data transmission system is responsible for the communication and data transmission among the stations. During the launch, the communication between the stations usually starts with calling the designators of each station, such as Huashan, Taishan, Yinhe, Yuanwang, etc. and Yuanwang-1 TT&C ship is shown in Fig. 5.4.

During the development of China's first artificial satellite, ground tracking stations were established all over the country. Xi'an, roughly in the center of China in geographical terms, was designated as the site for the command and control center, responsible for commanding the entire TT&C system. As the implementation of the manned space program unfolds, the Beijing Aerospace Control Center (BACC) was established in Beijing to take charge of the TT&C system of manned space program.

Due to the fact that the orbit injection often happens far away from China's territory, overseas tracking becomes necessary. Since the 1970s, China began to establish a fleet of TT&C ships and built the TT&C system. The TT&C ships are responsible for the tracking of injection point. At present, China has three TT&C ships.

Fig. 5.4 Yuanwang-1 TT&C ship

5.3.5 Ground Application System

Space engineering missions aim to realize specific objectives, such as satellite communications, weather forecasting, or scientific experiments. Therefore, it is necessary to set up mission application center. In space systems engineering, it is called the ground application system.

Since the downlinked application data from spacecraft far exceeds the downlinked engineering parameters data (housekeeping data) that are necessary to complete the TT&C, separate data receiving stations should be established, according to the different requirements of the missions, within the framework of ground application system.

Normally, the ground application system comprises a few subsystems, including data receiving and distribution, mission operation, and communication and data base.

Typical ground application centers in China are as follows: the China Centre For Resources Satellite Data and Application is responsible for the application of China's remote sensing satellites; the National Satellite Meteorological Center (NSMC) of China Meteorological Administration (CMA) is responsible for the application of China's Fengyun series meteorological satellites; the National Satellite Ocean Application Service is responsible for the data application of China's oceanographic satellites; the Technology and Engineering Center for Space Utilization, Chinese Academy of Sciences (CSU) is responsible for the application of China's manned space program; and the National Space Science Center (NSSC) of the Chinese Academy of Sciences is responsible for the application of China's space science satellites. Figure 5.5 shows the mission operation center at the National Space Science Center of the Chinese Academy of Sciences.

Fig. 5.5 The mission operation center located at the National Space Science Center, Chinese Academy of Sciences

Reference

1. Wertz JR, Larson WJ (1999) Space mission analysis and design, 3rd edn. Springer, Berlin

Chapter 6
Technical Fundamentals (I): Orbit, Attitude, and TT&C

6.1 Introduction

It is very important to grasp the fundamental knowledge related to spacecraft. Three chapters, including the present one, will be devoted to the technical foundations required by the implementation of space science experiments.

The spacecraft flying in space should follow the principles of orbit dynamics. Understanding these principles leads to better scientific objectives of space science missions and more reasonable technical specifications. Innovative scientific experiment concepts are also the outcome of the ingenious use of these principles. In a nutshell, the basic knowledge of orbit dynamics is essential for the proposers of scientific missions, as well as engineers and managers of space science missions.

A spacecraft that operates continuously in orbit needs to have a specific attitude, i.e., the relationship between its coordinate axis orientation and other directions of space, such as the center of the Earth or the Sun. For a free-flying spacecraft in orbit, its attitude is in the condition of inertial space of zero gravity. The attitude of a spacecraft in acceleration or deceleration is subject to the gravity in the acceleration or deceleration direction. Therefore, for the designers of space science experiments, it is necessary to acquire the fundamental knowledge of spacecraft attitude dynamics.

The ground measurement and control is closely related to orbit and attitude. This is also one of the most important duties of the TT&C system in space engineering.

6.2 Basic Concepts About Time and Position

First of all, a few basic concepts about time and position need to be clarified.

© Science Press 2021
J. Wu, *Introduction to Space Science*, Springer Aerospace Technology,
https://doi.org/10.1007/978-981-16-5751-1_6

6.2.1   About Position

The Earth rotates on its own axis with a period of 23 h 56 min 4 s (caution that it is not 24 h). This period is defined as sidereal day. At the same time, the Earth also orbits around the Sun. For a specific geographic location, when it returns to the same place heading to the Sun's direction, the Earth has traveled some distance and completed a full rotation. It takes about 4 min to return to the original alignment with the Sun. This period is defined as a solar day.

The exact value of a revolution period is 365.2422 days. A year is defined as 365 days by rounding off the number. The remaining 0.2422 day of each year is accumulated over 4 years and then 1 day is added to be the last day of February of that year. This year is defined as a leap year. The plane in which the Earth orbits about the Sun is called the ecliptic. The longest trajectory circumference of a point on the Earth created by the rotation of the Earth is defined as the equator. The angle between the ecliptic and the equator is the angle between the axis of Earth rotation and the normal of the ecliptic plane, which is about 23.5°.

The Earth rotation around the Sun creates four special astronomical positions, as shown in Fig. 6.1.

Vernal Equinox: The intersection between the ecliptic (the Sun's annual pathway from the South to the North) and the celestial equator.

Summer solstice: The farthest point north of the equator that the Sun can reach along the ecliptic (the Sun's annual pathway from the South to the North).

Autumnal equinox: The intersection between the ecliptic (the Sun's annual pathway from the North to the South) and the celestial equator.

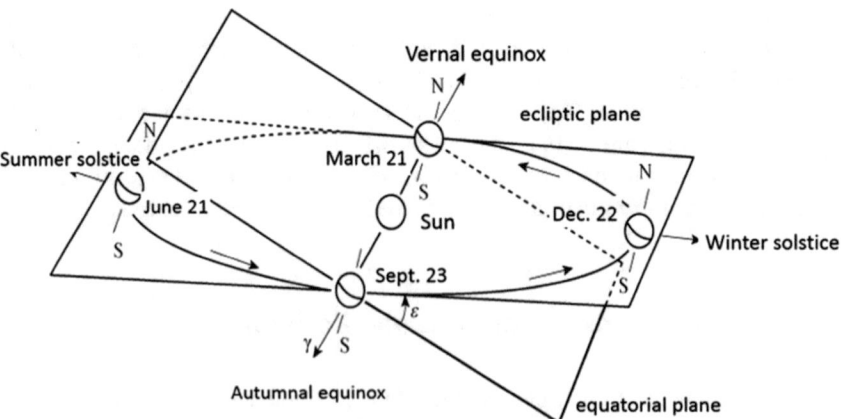

Fig. 6.1 Equator and ecliptic

Winter solstice: The farthest point south of the equator that the Sun can reach along the ecliptic (the Sun's annual pathway from the North to the South).

For the satellites that are orbiting the Earth, its center of the circular orbit must overlap with the Earth's center of mass so as to keep the circular orbit stable. As for a stable elliptical orbit, one of the focuses must overlap with the Earth's center of mass. For the interplanetary spacecraft that flies out of the Earth's gravitational field and reaches the second cosmic velocity, it will fly along a highly elliptical orbit, if no propulsion is applied. One of the focuses will coincide with the Sun's center of mass.

In order to quantitatively determine the position of a spacecraft in space, the following basic coordinate systems are defined.

The heliocentric ecliptic coordinate system: The origin is at the Sun's center of mass, the XY-plane coincides with the ecliptic plane, the X-axis points to the vernal equinox, and the Y-axis and Z-axis point to the direction of the Earth's orbiting around the Sun, determined by the right-hand rule (Fig. 6.2).

The geocentric equatorial rotating coordinate system: The origin is at the Earth's center of mass, the XY-plane is the equatorial plane, the X-axis along the equatorial plane intersects with the Greenwich meridian, the Z-axis points to the North Pole, and the Y-axis is determined by the right-hand rule. Since Earth is constantly rotating, this coordinate system is called the rotating coordinate system (Fig. 6.3).

Ground station coordinate system: The coordinate system of ground station on the Earth's surface is defined as that the XY-plane is the horizontal plane of the Earth, with the X-axis pointing due north and the Z-axis pointing to the zenith, and the Y-axis pointing due east as determined by the left-hand rule (Fig. 6.4).

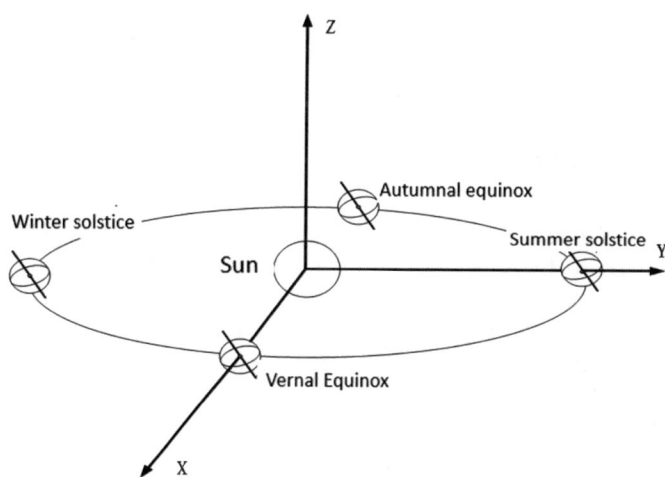

Fig. 6.2 The heliocentric ecliptic coordinate system

Fig. 6.3 The geocentric
equatorial rotation
coordinate system

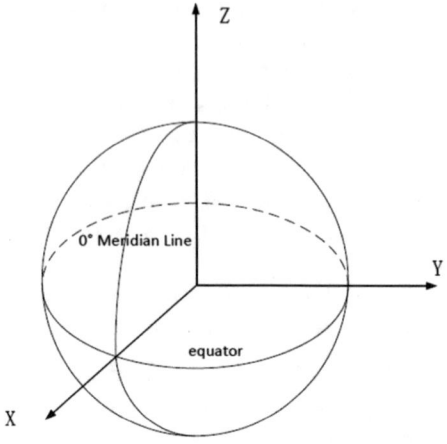

Fig. 6.4 Ground station
coordinate system

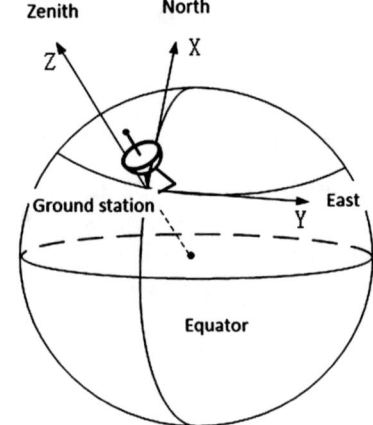

6.2.2   *About Time*

The accuracy of time is not only the basis of carrying out observations and experiments, but also the basis of space position measurement. For the implementation
of space systems engineering, each system has its own timing. If there is no unified
management and calibration of times, it is impossible to guarantee the mission execution and the accuracy of space position measurement. Therefore, it is necessary to
standardize the definition of time in advance. Table 6.1 shows several international
standardized times.

Table 6.1 Internationally standardized times

Times	Description
Solar time	The intervals between two consecutive alignments of a certain meridian circle with the center of the Sun (the local geographic time)
Universal time, UT	The mean solar time of the Greenwich meridian (0° longitude) where the Greenwich Observatory is located, also known as Greenwich time
UT0	It represents the initial values of Universal Time obtained by optical observations of star transits at various astronomical observatories across the globe. These values differ slightly from each other because of the effects of polar motion
UT1	UT1 is obtained by correcting UT0 for the effects of polar motion, the cause for meridian circle displacement
Atomic time, AT	A timescale based on the vibration frequency of electromagnetic waves generated by the internal motion of cesium atoms[6.1]
International atomic time (TAI)	It is based on the comparative study and processing of data from a system consisting of a large number of laboratory-constructed atomic clocks
Coordinated universal time, UTC	Because of the irregular slowing of Earth's rate of rotation by tidal friction and other forces, the UT has the tendency to slow down in the long run. *there is now about one more (atomic clock-derived) second in a solar year than there are UT1 seconds. To remedy this discrepancy, UTC is kept within 0.9 s of UT1 by adding a leap second to UTC in the mid of the year or at the end of the year. Therefore, UTC serves to accommodate the timekeeping differences that arise between atomic time (which is derived from atomic clocks) and solar time (which is derived from astronomical measurements of Earth's rotation on its axis relative to the Sun)
Local time	The time based on the local longitude
Beijing time	8 h earlier than universal time, which does not accurately represent the actual longitude of Beijing (116°20′E)

[6.1]*Note* It is a timescale generated by atomic clocks, which furnish time more accurately than was possible with previous astronomical means (measurements of the rotation of the Earth and its revolution about the Sun)

6.3 Fundamentals of Spacecraft Orbit Dynamics

6.3.1 Johannes Kepler's Three Major laws of Planetary Motion

(1) The planets move in elliptical orbits with the Sun at one focus.

Fig. 6.5 Kepler's second
law

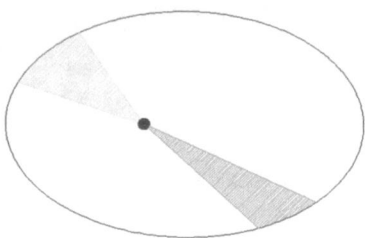

Table 6.2 The orbit
parameters of the eight
planets in the solar system

Planets	Orbital period/Earth year	Radii of the orbit	Alignment with Earth/Earth day
Earth	1	1	–
Mercury	0.24085	0.3871	115.88
Venus	0.61521	0.7233	583.92
Mars	1.881	1.5237	779.94
Jupiter	11.862	5.2028	398.88
Saturn	29.458	9.539	378.09
Uranus	84.015	19.19	369.66
Neptune	164.82	30.07	367.49

(2)   The time necessary to traverse any arc of a planetary orbit is proportional to
      the area of the sector between the central body and that arc (the "area law"),
      as illustrated in Fig. 6.5.
(3)   The ratio of cubes of the semi-major axis of planets' elliptical orbits to the
      squares of the planets' periodic times is a constant.

   Table 6.2 shows the orbit parameters of the eight planets in the solar system.

6.3.2   Spacecraft Orbit Dynamics

(1) Orbital Elements

The orbits of Earth satellites are described by six orbital elements, as shown in
Fig. 6.6, namely, semi-major axis a, eccentricity e, orbit inclination I, longitude of
the ascending node Ω, argument of perigee ω, and true anomaly θ. Among them, the
semi-major axis and eccentricity collectively determine the shape of the orbit. As for
the circular orbit, the eccentricity is close to 0 and the semi-major axis is the radius
of the circular orbit; as for elliptical orbit, the value of eccentricity e is between 0
and 1. The orbit inclination, right ascension of ascending node, and argument of
perigee are used to describe the position of the orbit. In Earth's coordinate system, a

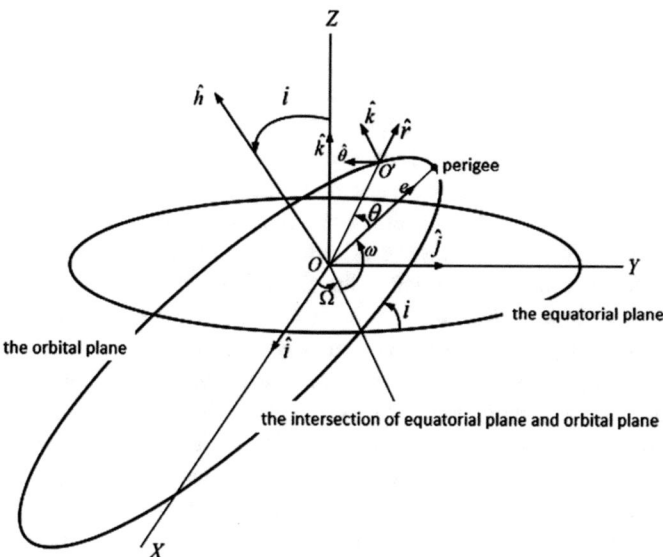

Fig. 6.6 The orbital elements of satellites

circular orbit or an elliptical orbit could be in different positions. The inclination of an orbital is defined as the angle between the orbital plane and the equatorial plane. If the orbital plane coincides with the equatorial plane, the inclination is zero; if the orbit is a polar orbit, the inclination is 90°. The longitude of the ascending node is the angle between the X-axis in the Earth-Centered Inertial Coordinate (namely, the intersections between the equator and the 0° meridian line) and the intersection line of the equator when the orbit ascends and the Earth center. The argument of perigee is the angle formed by the orbital perigee, ascending node, and the Earth center of the Earth. Once a satellite is launched, it will operate in inertial space, free from the rotation of the Earth. The position of the orbit in the inertial space will be determined by the launch time T_0. This factor will be qualified by the true anomaly which determines the position of the orbit in interstellar space.

(2) Orbit Dynamics

When a satellite is in orbit, even if it does not perform maneuvers and travels by inertia, it is still subject to the following perturbations that gradually change its orbit: perturbations caused by the Earth's gravitational inhomogeneity, perturbations caused by atmospheric drag, perturbations caused by the gravitational force of the Sun and the Moon, perturbations caused by solar radiation pressure, and perturbations caused by geomagnetism. The perturbations caused by attitude adjustment and orbital control air injection are active perturbations.

The mass distribution inside the planet Earth is not absolutely uniform. For a satellite in Near Earth Orbit (NEO), it is subject to weak perturbations of the changes in Earth's gravitational field. When the mass of the ground, toward which the satellite is heading, increases, the satellite's speed is slightly accelerated under the gravitational influence. In the same line of reasoning, after flying passed this region, the satellite's speed decelerates slightly due to the effect of gravity. In addition, the Earth is not perfectly round in shape as expected, but a bit oblate with a slightly larger radius at the equatorial plane, which also has an effect on the orbital velocity of the satellite.

Atmospheric drag also affects the speed of a satellite, and the effect is continuous. This effect is magnified by a short-term increase in atmospheric density during space weather events. Under the influence of atmospheric drag, the elliptical orbit will gradually become more circular, while the altitude of a circular orbit will gradually decrease. What needs particular attention is that this effect is directly related to the ratio of the satellite's windward area to its mass. In other words, if the satellite is large and light, the perturbation of atmospheric drag will be big, and vice versa.

The third-body gravitational forces of the Sun and the Moon also have an effect on the orbits of Earth satellites. The main third-body gravitational forces come from the Sun and the Moon. When the orbital altitude is less than 10,000 km, the third-body gravitational perturbations are relatively small. When the orbital altitude is higher than 50,000 km, the solar and lunar gravitational perturbations exceed that of the Earth's oblateness. The third-body gravity has an important impact on the long-term evolution of highly elliptical orbits and its lifetime.

Compared with the previous orbital perturbations, solar radiation pressure perturbation has tiny impact on the satellites in low Earth orbit.

Geomagnetic perturbations mainly apply to the satellites with fixed remanence magnetic field. Its effect on attitude cannot be ignored.

In addition, like all spinning objects, orbits have precession and nutation [1]. In particular, the precession deflects the normal direction of the orbital plane. As for orbit design, the deflection can equal to the daily rotation of the Earth around the Sun, so that the orbital plane is synchronized with the direction of the Sun. Such an orbit is called Sun-synchronous Orbit (SSO).

6.3.3 Examples of Commonly Used Orbits

(1) Sun Synchronous Orbit (SSO)

The orbital inclination is a little more than 90°, and the angular velocity of the precession of orbital plane is the same as the angular velocity of the Sun's movement on the equator. That is, the rotation from west to east is 0.9856° per day, and 360° per year. Therefore, the time of passage over the same geographic location for a certain latitude is the same. For Earth observation satellites, such an orbit ensures that the Sun's irradiation angle to the ground is the same for each pass, facilitating research that needs the same condition, as shown in Fig. 6.7. If the pass time is around noon, the shadow area will be the minimum. This is especially useful for urban remote sensing, since the maximum solar irradiation area can be obtained. For microwave remote sensing satellites that require a lot of solar energy, such as high-powered

Fig. 6.7 SSO

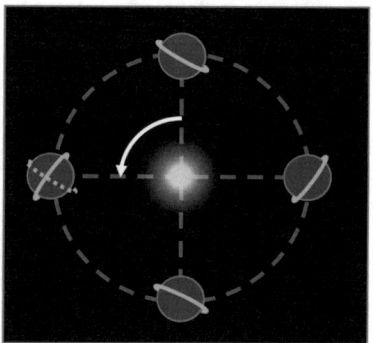

Synthetic Aperture Radar (SAR), and do not rely on the brightness of sunlight for imaging, a Sun-synchronous orbit with a dawn/dusk pass can be an option.

For scientific satellites, such as the solar observation missions, where the longest period of solar visibility is desired, a Sun-synchronous orbit with dawn/dusk pass is also an optimum option. For astronomical satellites, there is no constraints on Earth observation, and a Sun-synchronous orbit with dawn/dusk pass ensures stable satellite temperature parameters, making it easy to achieve stable spacecraft thermal control.

(2) Geosynchronous Earth Orbit (GEO)

The Geosynchronous Earth Orbit is above Earth's equator, as shown in Fig. 6.8, in which orbital period is equal to the Earth's rotation period, with zero eccentricity and zero inclination.

The satellite appears to be stationary to an observer on the Earth, with an orbital period of one sidereal day, or 23 h 56 min 4 s, and the altitude of satellite orbits is 35786 km. If three satellites are evenly distributed at 120° intervals, global coverage can be achieved, which is very desirable for applications in the fields of communication, broadcasting, meteorology, etc.

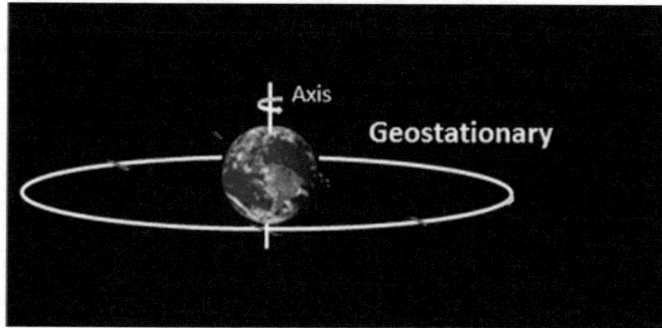

Fig. 6.8 Geosynchronous Earth orbit

Fig. 6.9 Elliptical orbit

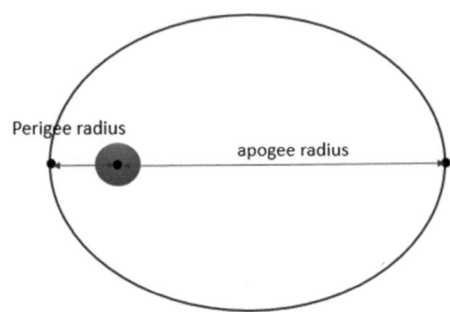

(3) Non-Sun-Synchronous Polar Orbit or Inclined Orbit

The orbit inclination is 0–90°, and the longitude of ascending node does not change in inertial space. However, due to the rotation of the Earth, the pass time of a satellite to a certain geographical location is constantly changing. Therefore, it is a suitable orbit for remote sensing satellites that need to observe the effects of different time periods and different solar elevation angle.

(4) Elliptical Orbit

According to Kepler's second law, the apogee of an elliptical orbit (Fig. 6.9) can be arranged close to the detection area by taking advantage of the relative slow motion of the satellite at apogee, which can increase the detection time in that area. However, due to the Earth's annual orbiting to the Sun, for the Earth, the orbit will change accordingly in inertial space. At the same time, the orbit itself and its apogee position will also produce precession. Therefore, when designing the orbit parameters, these factors need to be taken into account, and then the parameters should be further optimized.

(5) Molniya Orbit (Lightning Orbit)

There is an elliptical orbit of unique nature, namely, the Molniya orbit, also known as lightning orbit, named after the Soviet Molniya series communication satellites that started to use the orbit since 1960. This means that the orbital apogee does not change with time when the orbital inclination is 63.4 or 116.6°, and remains over the northern or southern hemisphere, as shown in Fig. 6.10. Although such an orbit is not always synchronized with the Earth's rotation like a synchronous orbit, it is suitable for the Earth observation satellites or regional communication satellites, because the satellite travels slowly at apogee and for most of the orbital period, it can cover the northern (or southern) high-latitude areas.

(6) Sun-Earth Lagrangian Orbit

The point, at which a satellite can remain relatively stationary because of the gravitational forces of the Sun and the Earth, is called a Lagrangian point. There are five L points, as shown in Fig. 6.11. The orbits around the Lagrangian point are Halo orbit (approximately circular) and Lissajous orbit.

Fig. 6.10 Molniya orbit

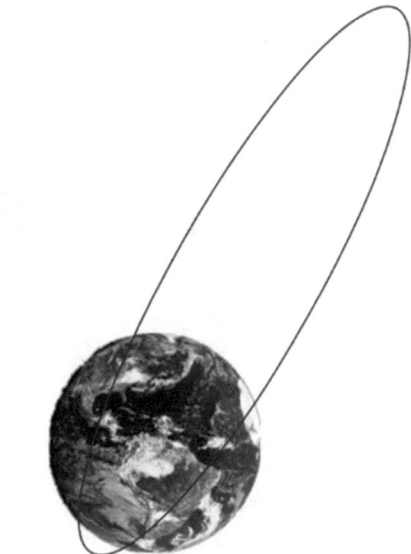

Fig. 6.11 L points, the points of gravitational equilibrium in the Sun-Earth-Moon system

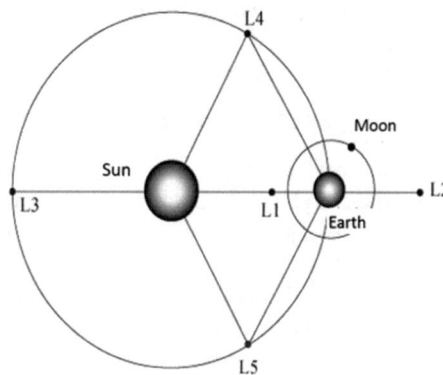

6.3.4 Orbit Maneuver and Limited Thrust

(I) Cosmic Velocity

The first cosmic velocity: the minimum launch velocity required for a satellite to be launched from ground and become an artificial Earth satellite, $V_1 = 7.9$ km/s.

The second cosmic velocity: the minimum launch velocity required for a satellite to be launched from ground and escape from Earth's gravitational field, $V_2 = 11.2$ km/s.

The third cosmic velocity: the minimum launch velocity required for a satellite to be launched from ground and escape from the solar system, $V_3 = 16.7$ km/s.

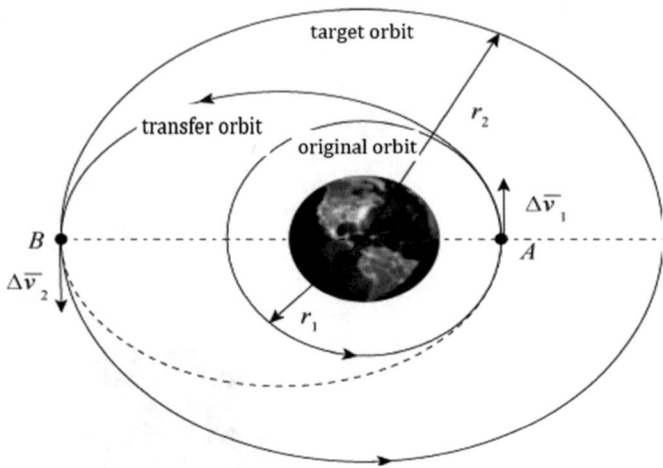

Fig. 6.12 Hohmann transfer orbit

(2) Hohmann Transfer Orbit

The best way to change the orbit between two coplanar circular orbits is by the Hohmann transfer orbit [2], as shown in Fig. 6.12. It is the best way for the spacecraft to go beyond the Earth orbits.

(3) Principles of Rocket Engine

Chemical fuels (including fuels and oxidizers) are burned in the combustion chamber to produce high-temperature, high-pressure gases, which are ejected at high speed through the engine Laval nozzle to produce reaction thrust, so that the rocket is accelerated.

Tsiolkovsky's formula[1]: The equation of conservation of momentum is based on the ideal condition without considering aerodynamic forces and the Earth's gravity, i.e., the increased momentum of the rocket is equal to the momentum of the gas ejected from the rocket, and the signs are reversed.

$$m\frac{dv}{dt} = -I_{sp}\frac{dm}{dt} \tag{6.1}$$

$$m_0 m_t e^{\frac{\Delta V}{I_{sp}}} \tag{6.2}$$

As above, m_0 is the mass before ejection; M_t is the mass after ejection; I_{sp} is specific impulse; and ΔV is for velocity increment. I_{sp} means the thrust produced per unit rate of consumption of the propellant. It is a measure of the efficiency of a

---

[1] *Note* Tsiolkovsky's formula provides the mathematical relationship between the changing mass of a rocket as it burns through fuel, the velocity of the exhaust gases, and the final speed of the rocket. It is considered as a foundation of astronautics.

rocket engine. As shown in Eq. (6.2), with the same mass consumption, the greater the specific impulse, the greater the ΔV.

Table 6.3 shows the relevant parameters and applications of rocket engine propellant.

Table 6.3 Relevant parameters and applications of rocket engine propellant

Types of engine	Propellant (fuels and oxidizer)	Specific impulse (s)	Features	Applications
Solid engine	Polyurethane, polydibutene, nitrate, plasticized polyether	approx. 285	Simple, reliable, and low cost	Apogee orbit change
Single-component liquid engine	Anhydrous hydrazine, hydrogen peroxide	approx. 200	Simple, low performance	No orbit maneuvers; for small satellite
Two-component liquid engine	Kerosene—liquid oxygen	approx. 320	High-performance complex system	Orbit maneuvers; orbit change
	Room temperature propellants dimethylhydrazine and nitrogen tetroxide	approx. 300		
	High-performance cryogenic propellant liquid hydrogen and liquid oxygen	approx. 420		
Hall electric propulsion	Xenon	approx. 1,600	Very high performance, very high power consumption, low thrust	Attitude and orbit adjustment
Ion electric propulsion	Xenon	approx. 3,000	Very high performance, very high power consumption, low thrust	Attitude and orbit adjustment
Cold gas propulsion	Nitrogen, helium, etc.	50–75	Simple, pollution free; very low performance	Pollution sensitive, for small satellites

6.4   Fundamentals of Satellite Altitude Dynamics

6.4.1   Commonly Used Altitude Stabilization Methods

(1) Simple Spin Stabilization

The satellite is directionally stabilized in inertial space due to the gyroscopic fixation obtained by the satellite rotation around its spin axis. The control is simple for the satellite spinning along the axis of maximum inertia. Most of the early satellites adopted this kind of control method. For example, the simple spin stabilization is applied on the Dongfanghong-1 (DFH-1, the first Chinese artificial satellite), and the TC-1 and TC-2 satellites of the Double Star Program (DSP). TC-1 and TC-2 satellites together with four satellites of ESA's Cluster Mission enabled the six-point detection of the geospace, the first time in human history. TC-1 is an equator orbit satellite in highly elliptical orbit with an orbital altitude of 550–74,017 km, while TC-2 is a polar orbit satellite in highly elliptical orbit with an orbital altitude of 700–35,000 km.

(2) Dual Spin Stabilization

Dual spin stabilization is realized by taking advantage of the fact that the satellite's spin is directionally stable in inertial space, and at the same time the payloads will remain stable by rotating with another speed, pointing toward the Earth or another object. For example, the remote sensor on-board the Fengyun-2 series (meteorological) satellites is on the main spin body, but the antenna that is responsible for the communication with the ground is slowly spinning.

(3) Momentum Bias Three-Axis Stabilization

The satellite has a large angular momentum in a certain axis of the main body axis (usually in the direction vertical to the orbital plane), and therefore there is gyroscopic stability in that axis, including fixed momentum bias stabilization, single-degree-of-freedom momentum bias stabilization, and double-degree-of-freedom momentum bias stabilization. It is suitable for the Earth observation satellites.

(4) Zero Momentum Three-Axis Stabilization

Each axis has an angular momentum, but the total angular momentum is zero, making the satellite control flexible with high pointing accuracy. It is ideal for various scientific observation satellites and suitable for all kinds of scientific observation satellites.

(5) Passive Stabilization of the Gravity Gradient

After the satellite enters orbit, the boom extends upward and it is relatively farther away from the Earth than the rest parts of the satellite, so the boom is subject to less gravitational force. The other end of the satellite is closer to the Earth and is subject to a bigger gravitational force. The gravitational difference forms a restoring torque for the satellite's center of mass. If the boom deviates from the local plumb line, this torque will allow it to restore its original attitude. There will be tendency for the axis of maximum inertia to be perpendicular to the orbital plane (pitch axis) and the axis of minimum inertia will be aligned with the local plumb line. This kind of control method is simple, practical, and durable, but the control accuracy is low (1–10°). It is only suitable for the Earth observation or communication satellites that do not require high accuracy.

6.4.2 Satellite Attitudes Description

(1) Attitudes Reference Frame

In order to quantitatively measure and control the satellite attitude, two coordinate systems, namely, the centroid orbit coordinate system and the ontology coordinate system, are usually used to describe the attitudes of satellite, as shown in Fig. 6.13.

I. Centroid Orbit Coordinate System $(OX_oY_oZ_o)$

The origin O is at the center of mass of the satellite, the orbital plane of the satellite is the coordinate plane, the Oz_o-axis points from the center of mass to the Earth center, the Ox_o-axis is perpendicular to the Oz_o-axis in the orbital plane and points to the direction of the satellite velocity, and the Oy_o-axis is parallel to the normal line of the

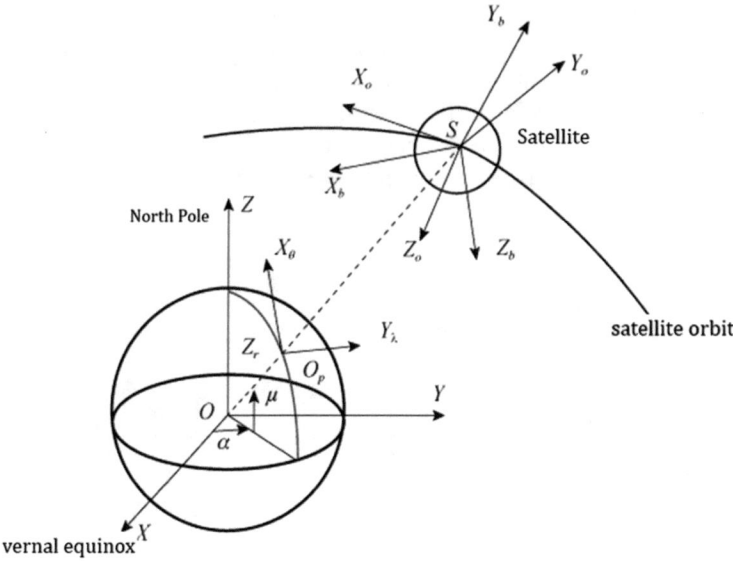

Fig. 6.13 Satellites attitudes reference frame

orbital plane. This coordinate system is suitable to define a three-axis stabilization Earth observation satellite.

II. Ontology Coordinate System (Ox_bY_bZ_b)

Origin O is at the satellite's center of mass, Ox_b points to the satellite top along the longitudinal axis of the satellite, Oy_b is perpendicular to the Ox_b-axis in the longitudinal symmetry plane of the satellite, and Ozb follows the right-hand rule. This coordinate system is suitable to define satellites with a distinct longitudinal axis.

(2) Changes in Satellite Attitude

Three Euler angles of rigid body rotation (i.e., pitch, yaw, and roll) can be used to define the changes in satellite attitudes.

In addition, the attitude changes caused by rotation can also be defined by the four elements of the attitude, i.e., the projection of the unit vector of the rotation axis in the reference coordinate system as described by the three directional cosines $e_x e_y e_z$, and the angle of rotation Φ around this axis.

The attitude changes of the spin-stabilized satellite can be described by the orientation of the spin axis in the equatorial inertial coordinate system (right ascension and declination), as well as the angular velocity of the spin.

6.4.3 Satellite Attitude Control

The satellite attitude control system is composed of attitude sensor (attitude measurement component), attitude actuator, and attitude algorithm, as shown in Fig. 6.14. Table 6.4 shows the measurement accuracy and characteristics of on-board sensors.

The common actuators for attitude control are flywheels, jet actuators (micro-jet engines or plasma engines), or magnetic torquer. Different actuators are selected according to the attitude adjustment requirements, as well as the satellite mass and

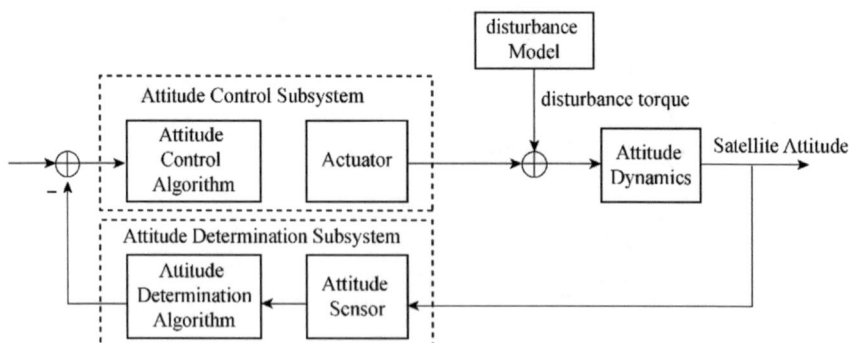

Fig. 6.14 Satellite attitude control system

Table 6.4 Measurement accuracy and characteristics of on-board sensors

Sensor	Accuracy of measurement	Characteristics and usability
Magnetometer	>10°@5000 km, >5°@200 km	Measures the attitude relative to Earth's local magnetic field and its uncertainty and time variability limit the accuracy. It is suitable for orbital altitudes below 6000 km
Earth sensor	0.05°@geostationary orbit, 0.1°@low Earth orbit	The uncertainty of the horizon limits its accuracy. The scanning method is generally used for high-precision Earth sensors
Sun sensor	0.01°	The normal field of view is ±30°
Star sensor	2″	The normal field of view is ±6°
Gyroscope	0.001°/h	The drift of reference position should be corrected periodically during the normal use
Radio frequency sensor	0.01–0.5°	Generally applied to antenna capture and tracking

moment of inertia. Among them, flywheel and magnetic moment converters both consume only renewable electrical energy, while jet engine and plasma engine both consume substance to varying degrees and therefore have lifetime issues.

6.5 Tracking, Telemetry, and Control (TT&C)

6.5.1 The Responsibilities of the TT&C System

On the ground, it is necessary to obtain the real-time information pertaining the satellites' orbit position and attitude, which are subject to necessary adjustments according to the mission requirements. This task is undertaken by the ground Tracking, Telemetry, and Control system.

The foremost task after the launch is the telemetry which will collect the data of the working conditions of various instruments, engineering parameters, environmental parameters, and other relevant data, and send the data to the ground TT&C system through the satellite-ground downlink in real time or in a delayed manner, so as to monitor the satellite from the ground.

The process of sending the mission command to the satellite is called remote control where the satellite will receive the remote control command (injected data) or programming sent by the ground stations through the ground-satellite uplink, and then transmit it to the on-board instruments for execution to realize the satellite control.

Orbit tracking and measurement is also the main task of the TT&C system. Multiple ground stations will, in coordinated manner, send radio signals to satellite, which will be returned by the on-board transponder so as to measure the satellite's velocity, distance, angle (azimuth and pitch), and the satellite orbit parameters will be eventually calculated on the ground.

When the data traffic is heavy, the satellite needs to be equipped with a dedicated data communication link for the payloads, and the data will be sent to the ground data receiving station through the satellite-ground downlink, which will be handled and processed into application data products.

6.5.2 Technical System of TT&C System

At present, the technical systems of TT&C have been standardized internationally, which are introduced as follows.

(1) United Carrier TT&C System

Multifunctional synthesis of TT&C is realized by using one principle carrier with multiple subcarriers (PM system), and each subcarrier carries one kind of data with multi-side tone ranging, such as USB tracking telemetering and control system with uplink 2025–2110 MHz, downlink 2200–2290 MHz. Telemetry: 8 kbps, PCM-DPSK-PM, subcarrier 65.536 kHz. Remote control: 2 kbps, PCM-PSK-PM, subcarrier 8 kHz. Measurement track: 7 side tones.

(2) Spread Spectrum United Tracking Telemetering and Control System

Pseudocode realizes multi-functional TT&C, data time-division multiplexing (TDM), pseudocode ranging, and support single-carrier multi-target TT&C.

(3) Deep space United Tracking Telemetering and Control System

Residual carrier modulation, TURBO code, and Very Long Baseline Interferometry (VLBI) goniometric signal. The Chinese VLBI Network (CVN) affiliated to Chinese Academy of Sciences consists of four antennas in Beijing, Shanghai, Kunming, and Urumqi, and a data processing center at Shanghai Observatory.

6.5.3 Chinese TT&C Network

China's TT&C network is responsible for the execution of state-approved space missions and a certain number of commercial space missions. The network consists of Beijing Aerospace Control Center, China Xi'an Satellite Control Center, a number of ground stations scattered around the country, the TT&C ships, a relay satellites network in geostationary orbit, as well as a relay satellite at L2 point. The network has the capacity of providing global coverage for the Earth orbit spacecraft as well as solar system probes flying as far as to the moon.

6.5.4 Satellite Tracking and Methods of Orbit Measurement and Determination

The commonly used orbit measurement and orbit determination methods are as follows.

(1) Single Station Measurement

The oblique distance from the satellite to the radar station can be measured by single radar station with monopulse radar; the rate of change of the slant ranges can be measured by Doppler frequency shift; the azimuth and elevation angle of the satellite relative to the radar station can be measured by the radar antenna turntable; and the rate of change of azimuth and elevation angle can be measured by the radar antenna turntable tracking system. Therefore, six independent parameters in total, after repeated measurements, can determine the orbit.

(2) Three-Station Measurement

In the case of three ground stations, the orbit measurement facility only has the range information, without information of velocity and angle measurements, the slant ranges from the satellite to the ground stations can be measured at the same time to get the position vector of the satellite at a particular time. The orbit can be determined after repeated measurements.

(3) VLBI Orbit Measurement

The VLBI array on the ground, the main facility for radio astronomical observation, can be used to measure and determine the orbits beyond the Earth orbits with higher precision.

(4) Orbit Calculation and Determination

Without considering the influence of the perturbative force on the orbit, the orbit that is calculated on the basis of preliminary orbit measurement data is called the *initial orbit*. The error of initial orbit is generally big, but it can be used as the initial condition for more precise orbit determination or as the forecast for the ground stations to track the satellite.

The orbit measurement with longer time period can greatly improve the precision of the orbit. These orbit measurements, combining with a precise orbit dynamics model and proper estimation algorithm, can improve the orbit determination precision. This orbit is called improved orbit or orbit estimation. The algorithms for orbit estimation include: (i) least squares, a post-processing method with good convergence and high iteration accuracy; (ii) Kalman filter, a real-time processing method of orbit observation, but the convergence of the iterative operation must be ensured.

References

1. Lin L (2019) Satellite orbital mechanics algorithm. Nanjing University Press, Nanjing
2. Montenbruck O, Gill E (2000) Satellite orbits-models, methods and applications. Springer

Chapter 7
Technical Fundamentals (II): Scientific Payloads and Its Application Environment

7.1 Introduction

Chapter 5 briefly introduces eight subsystems of the satellite/spacecraft system, all of which are essential subsystems for the whole mission. In other words, these eight subsystems are integral parts of all satellites, such as meteorological satellites, remote sensing satellites, and space science satellites. The technical aspects of these eight subsystems are often collectively called the *satellite platform*.

In addition, if a satellite is to accomplish specific objectives, it is necessary to be equipped with instruments or scientific experiment packages which are called payloads in a satellite system [1]. Therefore, the ninth subsystem is the payload system. Depending on the complexity of specific missions, the payload system is likely to be composed of multiple instruments rather than one instrument. In most cases, the mission application department will lead to the development of payloads with the participation of multiple institutes. The payload system itself, more often than not, will be divided into a hierarchy of subsystems according to the different functions of the instruments. This hierarchy is useful for strengthening the systematic management of payloads, raising awareness of their importance in the whole satellite system, and recognizing their distinctive mission functions.

The on-board payloads are sometimes larger and heavier than the satellite platform, such as the Hubble Space Telescope, which appears to be a telescope rather than a satellite because the design of all subsystems related to the satellite platform focuses on the requirements of science payloads. However, in most cases, the science payloads are required to adjust to the general design constraints laid out by the satellite platform in terms of size, weight, power supply, mounting surface, thermal environment conditions, mechanical environment conditions, radiation environment conditions, etc. These conditions are crucial input for the technical design of science payloads. If the science payloads are piggybacked with an application satellite, they must meet the mission's design requirements, leaving the science payloads little freedom to ask for individually tailored conditions. In the case that the science payloads are the mission's primary payloads and the platform cannot

© Science Press 2021
J. Wu, *Introduction to Space Science*, Springer Aerospace Technology,
https://doi.org/10.1007/978-981-16-5751-1_7

provide satisfactory conditions, special requirements can be put forward, especially when these conditions determine the realization of science objectives, such as attitude stability, special temperature environments. The satellite platform leaves no stone unturned to create these conditions for the science payloads or provide compromised options, especially when materializing these special requirements the payloads themselves will make the mission development more economically feasible, such as more precise pointing control or simply ensuring a lower temperature in the vicinity of the detectors.

Compared with the application satellites, space science satellites that have different orbit requirements may use orbits of different altitudes, periods, inclinations, and even orbits beyond the Earth orbits, which means the amount of space radiation that the science satellites may be exposed to during their lifetime is also very different from that of application satellites. Therefore, it is necessary to give space radiation special consideration in the design phase, and additional protection measures should be taken when necessary.

This chapter will first elaborate on the payloads of space science missions and then briefly introduce the environment requirements for the on-board science payloads.

7.2   Space Science and Science Payloads

Science payloads are the main aspects of a space science mission. The science payloads vary greatly from mission to mission, depending on the science objectives of a particular mission. Even the objective is the detection of the same physical parameters using the same detector, the detector itself may be updated to obtain physical parameters with higher precision. Therefore, it is difficult to provide a comprehensive overview of all space science payloads. This section will provide a general introduction of all types of payloads, which will be classified according to the electromagnetic spectrum and follow the order of low frequency to high frequency, e.g., from electrostatic, low frequency, microwave, millimeter-wave, terahertz, infrared, visible light, all the way to ultraviolet, X-rays, gamma rays particles.

7.2.1   Electrostatic Field, Magnetostatic Field, and Low-Frequency Electromagnetic Wave Detectors

Parameters of electric field and magnetic field always appear at the same time. According to Maxwell's equations, a changing magnetic field generates an electric field, and a changing electric field generates a magnetic field, making it possible, by measuring the magnetic field, to obtain the parameters of the electric field in the low-frequency band of electromagnetism, from sub-Hertz to tens or hundreds of Hertz. The reason behind this is that the low-frequency magnetic field is easier to measure than the low-frequency electric field. Such a detector is called a *low-frequency elec-tromagnetic wave detector*, and its probe is usually positioned several meters away from the satellite platform so as to use multiple coils to induce a changing magnetic field. The generated induced current in the coils will be amplified by a preamplifier and transmitted to the data handling and recording equipment located on the satellite.

However, the electrostatic field and the magnetostatic field are independent of each other and can only be detected separately. The electrostatic field is detected by measuring the electromotive force between two electrodes with a fixed distance, e.g., the detection of electric potential between the satellite and a charge collection region area at the top of the probe. The boom that extends from the satellite with a collection device (usually a spherical conductor) is called the Langmuir probe, as shown in Fig. 7.1. The Langmuir probe collects electric charges in space, and creates, with the satellite itself, an electromotive force, which will be converted into electrical signals. The signals will be amplified, recorded, and stored. Since there are few space charges, the probe is easily affected by the charges of the satellite itself. Therefore, it is necessary to take measures to make sure the satellite surface is electrically conductive and prevent the scenario that additional interfering charges gather in a certain area.

The detection of magnetostatic fields is relatively easier. Similar to the low-frequency electromagnetic wave detectors, a probe is required to be mounted at the top end of a long boom that is usually several meters in length so as to keep some distance from the satellite. The probe will work through a coil or other magnetic inductor, which is then amplified and transmitted to an Electric Box located on the satellite. However, in order to avoid interference from residual magnetic fields of various equipment and metal components on the satellite, which affect the detection

Fig. 7.1 Langmuir probe on-board the ESA Rosetta mission. Photo credit: ESA

of real space ambient magnetic field, it is also necessary to counteract the interference from residual magnetic field on the satellite by installing two probes at different positions of the same boom. If there is only the ambient magnetic field, the measurement results of both probes must be the same. If there is interference from the satellite itself, the probe closer to the main structure of the satellite will produce data of higher value.

For satellites that perform magnetostatic field measurements, electrostatic field measurements, and low-frequency electromagnetic field measurements, it is necessary to take measures of remnant magnetization suppression and surface electric potential control on the main structure of the satellite, which should be carefully considered at the initial spacecraft design phase and the remnant magnetization in the components and parts should be strictly controlled. Furthermore, testing of the remnant magnetization should be carried out before the factory acceptance. As for the surface electric potential control, a conductive film should be applied to the satellite surface.

7.2.2   Low-Frequency Radio Sensor

In the radio band (kHz to MHz), an antenna can be used as a probe to detect the electromagnetic field in space. According to the antenna radiation theory, when the physical size of the antenna is much smaller than the wavelength of electromagnetic waves, the antenna is called electrical small antenna, and its radiation (reception) efficiency will be very low. The impedance of the antenna to the feeder port changes dramatically, making the match difficult.

However, with the advancement of digital technology, many receiving devices with smart matching capacity continuously emerge, which can partially offset the loss of antenna receiving efficiency, making it possible to enable low-frequency electromagnetic wave detection by using electrically small antennas. The low-frequency electromagnetic wave detector on-board the TC-2 satellite of Double Star Program is shown in Fig. 7.2. When implementing the detection of electromagnetic waves below 50 MHz, consideration should be given to the shield of the ionosphere, the lighting in the terrestrial space as well as various interferences of industrial electromagnetic waves generated by human beings. In the low-frequency band, the far side of the moon is the most ideal place in space for the detection of low-frequency electromagnetic waves in the universe.

Due to the extremely wide beamwidth of the electrical small antenna pattern, to achieve effective, high spatial resolution directional observations, and imaging, it is necessary to form multiple small antennas array or form an interferometric baseline for measurement. This brings about the requirements of small satellite formations for scientific exploration, which is also the new direction of technological development in recent years.

Fig. 7.2 The low-frequency electromagnetic wave detector on-board the TC-2 satellite of Double Star Program

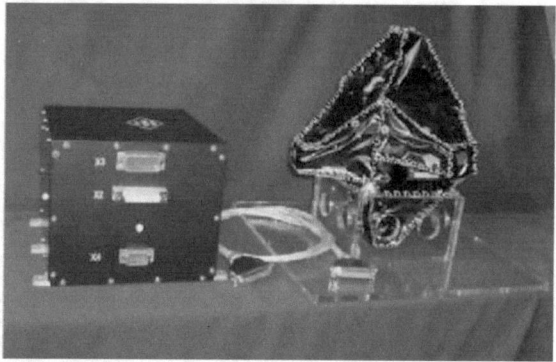

7.2.3 Microwave Remote Sensor

For radio frequency in the microwave band (1 G–30 GHz), the wavelength is reduced from tens of centimeters all the way down to several millimeters. Therefore, the size of the antenna unfolded on the spacecraft can reach the best radiation conditions (>1/2 wavelength), making possible the high-resolution Earth observation and observation of the universe. From the 1970s, starting with the first satellite with on-board microwave remote sensor, due to the all-day (day and night) and all-weather (under all kinds of weather conditions) advantages, microwave Earth observation has experienced rapid development in the course of 50 years. In particular, the advantage of a stable microwave signal source can be used, and the singles reflected from the ground can be recorded to mark the positions of the satellites. The recorded data will be processed after a few kilometers' flight, and in this way, a virtual antenna array on a kilometer scale will be formed. This technology is known as synthetic aperture radar (SAR) technology, whose spatial resolution is comparable to that of a visible light remote sensor. Figure 7.3 shows a multi-modal microwave remote sensor, the main payload on-board Shenzhou IV spaceship.

However, for passive microwave remote sensors that can only receive but not transmit electromagnetic waves, the realization of a synthetic aperture can only depend on real-time long-baseline interference and multi-baseline array. For the time being, the technical difficulty of high-resolution passive microwave observation with more than 20 m baseline by NEO satellites has not been solved yet.

Due to the fact that the wavelengths from microwave to millimeter-wave are ideal for water observation, microwave is the best frequency band for the observation of ocean, soil moisture, snow and ice, rivers and lakes, vegetation water content, atmospheric temperature, and humidity.

Fig. 7.3 Multimode microwave remote sensor on-board Shenzhou IV spaceship

7.2.4   Millimeter-Wave and Submillimeter-Wave Remote Sensor

In the millimeter-wave and submillimeter-wave bands, the atmosphere absorption occurs at some specific frequency bands (as shown in Fig. 1.1 in Chap. 1), which blocks the Earth observation from space, but it brings the information of the atmosphere, such as temperature and water vapor. Therefore, this wave band is the best frequency band for observing the troposphere and even the stratosphere. The antennas of millimeter-wave and submillimeter-band remote sensors are smaller and more efficient, but they also require higher processing precision, making them more costly.

In this frequency band, the efficiency of the components will gradually decrease and the noise will increase, making the overall payload development more difficult and costly. However, the basic design principle is essentially the same as that of microwave detectors.

Figure 7.4 shows the Geostationary Interferometric Microwave Sounder (GIMS) in millimeter-wave and submillimeter-wave band, and the high-resolution passive microwave images it produced. The GIMS is developed by the National Space Science Center of the Chinese Academy of Sciences, using interferometric imaging technology, for China's meteorological satellites in Geosynchronous Earth Orbit.

7.2.5   Terahertz Remote Sensor

Terahertz is a frequency band between submillimeter-wave and infrared, with a wavelength of 3–0.03 mm. It is a transitional spectral band of electromagnetic wave energy

Fig. 7.4 GIMS and the high-resolution passive microwave images it produced

from electron to particle (photon). There are new advancements in terahertz technology in recent years, but it is still an uphill battle for space probe development. In view that the atmosphere literally blocks the electromagnetic waves in this band, there are few applications of this band in Earth observation. However, due to its small size, it may be used in the future for interplanetary communication and detection of some special gas in the atmosphere of the planet Earth and other planets.

7.2.6   Infrared Remote Sensors

Infrared is the main frequency band of thermal radiation and there are extensive applications of this band. The on-board sensor is mainly composed of an optical telescope and infrared detector. The telescope part is similar to the visible telescope. Therefore, the infrared detector part is the key technology. In order to improve the detection sensitivity, it is necessary to take measures to cool the detector, such as space radiation cooling, Stirling mechanical cooling, and liquid nitrogen/liquid helium gas cooling. The detector cooling is a major technical difficulty in the development phase, as well as a major constraint on the lifetime of the remote sensor, which also increases the size and weight of the whole payload, and correspondingly increases the power consumption. In addition, because of its application in the national defense, the infrared detector and its key technologies are developed in a self-reliant manner and cannot rely on technology import. Currently, China already has the capability to develop infrared remote sensing technology independently. However, in terms of sensitivity, there are still gaps to fill compared with the most advanced level in the world. This is especially the case for scientific applications where there is a very high requirement for the sensitivity of infrared detection of the universe, making the most sensitive infrared detector in high demand. Figure 7.5 shows the infrared imaging spectrometer on-board Chang'e-4 lunar mission.

7.2.7   *Visible Light Remote Sensor*

The visible light remote sensor is the earliest on-board payload and it is the main remote sensor for Earth observation. However, the meteorological conditions and sunlight radiation conditions have constrained, to some degree, the application of visible light remote sensing. The visible light remote sensor is composed of two parts: the telescope and the imager. The telescope's optical system is similar to that of ground-based telescopes. The difference lies in that the aperture is increased by reducing weight and there are corresponding problems to increase temperature stability, which has evolved as the main technical development direction. The primary mirror of James Webb Space Telescope (JWST), the U.S. second-generation large space telescope, has reached 6.5 m in diameter.

Detector technology and the development of ground digital cameras are mutually enabling. In terms of digital imaging, the Hubble Space Telescope (HST), launched in 1990, has contributed to the development of charge-coupled device (CCD) detectors. In 2000, the Jet Propulsion Laboratory (JPL) developed the complementary metal-oxide semiconductor (CMOS) technology characterized by lightweight and low cost. The extensive use of CMOS detectors in ground cameras led to a rapid reduction in cost and a significant increase in signal-to-noise ratio. Currently, both CCD and CMOS are being used in a large number of space applications, with the cost of CMOS decreasing rapidly and CCD applied and extended to the UV and X bands. Figure 7.6 shows the medium-resolution wide-swath camera on the Gaofen-6 satellite.

Fig. 7.6   The medium-resolution wide-swath camera on the Gaofen-6 satellite. *Note* *Gaofen means "high resolution" in the Chinese language. The satellite is a Chinese civilian remote sensing satellite

7.2.8   Ultraviolet Remote Sensor

In the ultraviolet band, the wavelength is further reduced to hundreds of nanometers to tens of nanometers. The design, processing, and assembling of the optical system of the telescope have become increasingly difficult, and the suppression of the stray light has become a technical difficulty. With the development of CCD and CMOS technology, the UV remote sensor can achieve high-resolution imaging as in the visible band. From the ultraviolet to higher frequency bands, the spacecraft is the only observation platform, because it is no longer possible to carry out observations on the ground. Figure 7.7 shows the extreme ultraviolet camera on the Chang'e-3 lunar mission.

7.2.9   X-ray Remote Sensor

In the X-ray band, the energy of photons further increases up to hundreds of electron volts (eV). When the photons reach the optical reflector, they will easily pass through the medium or will simply get lost in the medium, making reflection impossible. Therefore, the entire optical system is designed as a small-angle grazing imaging system on a reflective surface coated with smooth metal. Consequently, there are two design options: a multi-layer barrel-type parabolic or hyperbolic reflector (Wolter I), and a multi-channel lobster-eye type reflector, as shown in Fig. 7.8. The detector can still use the expanded CCD or CMOS, but its efficiency will be significantly reduced.

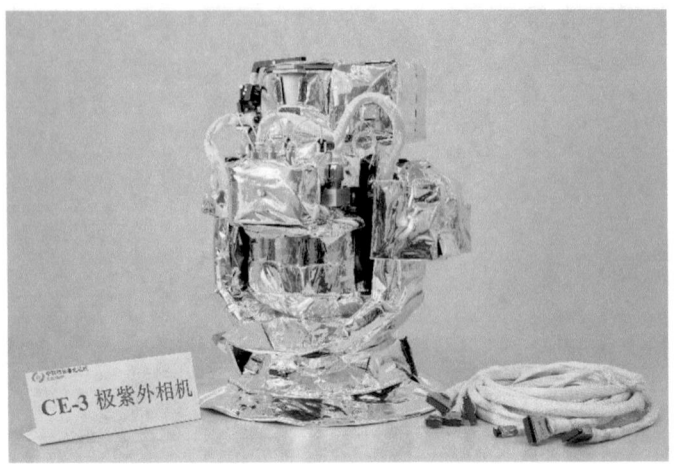

Fig. 7.7 The extreme ultraviolet camera on the Chang'e-3 lunar mission

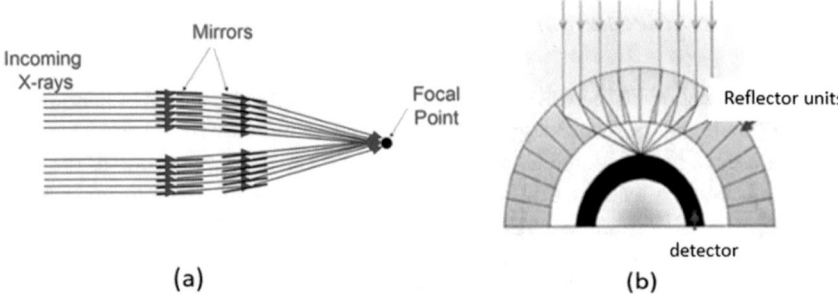

Fig. 7.8 Focused imaging: Wolter I type and lobster-eye type

7.2.9.1 Gamma Ray Detectors

Gamma rays in the universe are generated directly by their sources, including the Sun, with energy up to MeV, which is highly penetrating. They are impossible to be focused and detected directly by conventional emission and grazing telescopes, but they can be detected indirectly by the fluorescence excited by scintillating crystalline matter after being irradiated by gamma particles. The detector consists of a scintillation crystal, a photomultiplier, an amplifier, as well as an electronics box. Figure 7.9 shows the detector diagram of the DAMPE (DArk Matter Particle Explorer) satellite.

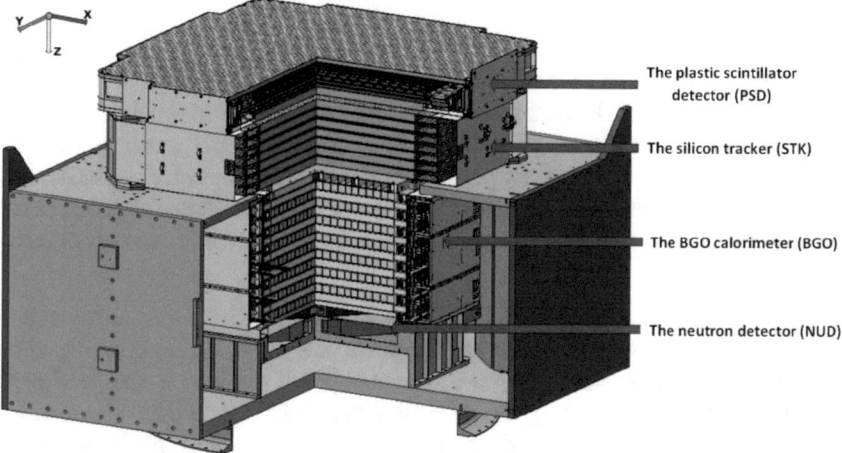

Fig. 7.9 The detector diagram of the DAMPE satellite

7.2.9.2   Electron and Particle Detectors

Compared with the detection of photons, the method to detect neutral particles, such as electrons and ions, and other neutral non-photon particles is very different. First, the energy distribution of these particles covers a wide spectrum. The energy spectrum of particles that scientific research always focuses on can range from a few electron volts (eV) to giga-electron volts (GeV) and even teraelectron volts (TeV).

For positively and negatively charged electrons and ions, the energy spectrum can be separated by deflecting them with a fixed magnetic field or a high-voltage electric field, while for neutral particles, it is necessary to resort to different ways, e.g., their mass and energy can be detected by measuring their time of flight between the two polar plates in the detector. Therefore, for particle detectors, there are many variations in types and designs. Figure 7.10 shows the high-energy proton-electron detector on-board the Fengyun-4 meteorological satellite.

Fig. 7.10 The high-energy
proton-electron detector
on-board the Fengyun-4
meteorological satellite

7.2.9.3   Utility Equipment

In the scenario that each payload individually has an electrical interface to the satellite platform, the cost is high and resources consumption is substantial. This is especially the case when there are many payloads involved. Therefore, utility equipment can be considered to provide supporting service to all the science payloads, e.g., providing the power supply, command, data acquisition, etc. Then, all the payloads are connected to the satellite via the utility equipment. The common types of utility equipment include:

(1)   Power distributor. It obtains the primary power supply +27 V from the satellite platform, and then produces various power supply voltages according to the needs of each payload, and provides switch on/off services. If a payload needs a special power system, such as high-voltage power for particle detectors, it is necessary to design a specific power system for the payload for its own sake.

(2)   Data manager. It provides various information services for the payloads, such as clock pulse, command transmission, engineering parameters acquisition, and scientific data acquisition, and coordinates the work progress of other individual instruments managed by the utility equipment. For example, the Tiangong-2 space laboratory incorporates 51 payloads and has conducted more than 10 scientific experiments, which are eventually managed by a unified data manager.

(3)   Mass storage. It is responsible for storing all kinds of data, especially a large amount of scientific data before it is downlinked to the ground station. It also has the capacity to package the required scientific data on the occasion of pass downlink. For example, the memory capacity of the Double Star Program is tens of megabytes (MB), and nowadays the capacity has rocketed to a gigabyte (GB) for the current satellites.

(4)   Digital transmitter. It is responsible for transmitting scientific data to ground stations. In terms of data transmission, meteorological satellites, communication satellites, and scientific satellites all have distinctive features. The TT&C system cannot provide data transmission for all types of satellites. Therefore, scientific satellites need dedicated data transmitters, independent of the satellite platform, to provide service to payloads.

With the development of electronic chip technology, the power distributor, data manager, and mass storage can be integrated together at the design phase so as to reduce weight and power consumption.

7.3   Satellite's Environmental Requirements for the Science Payloads

Although in most cases the design of space science satellites should primarily meet the needs of the science payloads, it is still the first choice to keep the technical specifications of the satellite platform unchanged or with minimum changes, because of the increasing maturity and reliability of satellite technology. This choice not only improves the mission reliability but also significantly reduces costs and increases the effectiveness of the national investment in space science. Therefore, when designing scientific payloads, it is necessary to lay the groundwork by getting familiar with the requirements of satellite platforms/bus for payloads, and on the condition that the scientific objectives can be well catered, excessive requirement beyond the current platform capacity should be avoided so as not to add the additional workflow for the mission design and development, and increase the unnecessary cost.

7.3.1 Mechanical Environment Requirements

The satellite platform usually has a mounting surface (s) for the payloads. If the payloads' mission is Earth observation or sky survey, the satellite platform will also have one or more sides to mount the payloads or its detection window. Solar panels are usually mounted on two symmetrical sides or just on one side of a satellite with three-axis attitude stabilization. In addition, there is a special surface, often referred to as the *bottom surface*, which will be connected to the launch vehicle. This means that a docking ring of about 1 m in diameter will be mounted to the bottom surface. After the satellite is separated from the rocket, this surface can be used for scientific observations, such as detecting particles from this direction. However, because it should be docked with the launch vehicle, this surface cannot install bulging instruments.

When the payloads are mounted on the mounting surfaces and plates both inside and outside the satellite, the mounting screws are needed to fix the mounting interface. For better thermal control efficiency, the mounting interface needs to have good contact with the instrument. Therefore, there is a high requirement for the flatness of the mounting surface, and the error usually cannot exceed a few microns.

Once all the instruments are mounted in place, the structural mechanics of the satellite need to meet the vibration mechanics requirements of the launch vehicle. The key characteristic frequencies in this requirement often fall in the low-frequency range from 10+ hertz to tens of hertz. Mechanical vibrations at higher frequencies normally come from the acoustic waves at launch, therefore not of significant magnitude. Once the characteristic frequencies of the satellite fall within the domain that is required to be avoided by the launcher system, structural adjustments of the satellite must be done, including adjustments to the payload's mounting position. Changing the characteristic frequencies will improve the mechanical environment to be experienced during the launch.

Depending on the requirements of the launch vehicle, the satellite platform will put forward mechanical test requirements for each instrument unit. Therefore, before delivery, each instrument unit should undergo mechanical experiments, which usually include random vibrations and sinusoidal vibrations, and the vibration tests will be performed in the X, Y, and Z directions. As shown in Fig. 7.11, the Quantum

Fig. 7.11 The QUESS satellite in mechanical vibration tests

Experiment at Space Scale (QUESS) satellite, also known as Micius, is conducting mechanical vibration tests.

When the satellite is separated from the launch vehicle, it is subjected to a large impact force. Therefore, during the development of the satellite, it is required to conduct a separation impact test. However, the force will be decomposed and the impact force reaching the payload's mounting position will be much reduced. In general, there will be no special requirements of mechanical conditions for payloads.

7.3.2 Thermal Environment Requirements

If the science payloads have special temperature requirements, this should be put forward at the early design phase, which will be considered according to the parameters of the satellite platform. If there is a requirement for the temperature of a payload, the requirement can be met by installing additional heat pipes under the instrument unit or by coating, e.g., a black coating is conducive to heat absorption, while the white coating is good for heat insulation. If the detector inside a payload needs local temperature control, the satellite platform can also provide a separate heat pipe to reach the specific local position that needs thermal control. Alternatively, the power supply requirement can be proposed for a specific payload and its own temperature control components can be added accordingly. Figure 7.12 shows the ultra-low-cooling technology approach applied on the single-photon detectors on

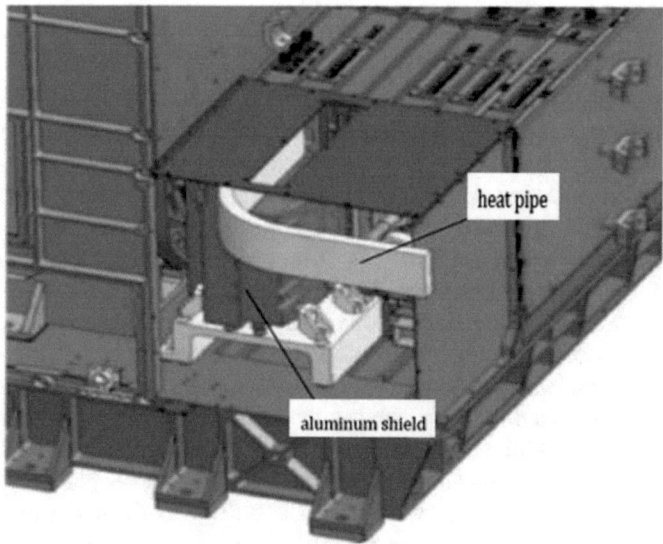

Fig. 7.12 The ultra-low cooling technology approach applied on the single-photon detectors on the QUESS satellite

the QUESS satellite. For both the three-axis stabilization attitude and the simple spin stabilization attitude, the satellite platform may have one of its sides facing the cold space for a long period of time, which is the best cooling environment. Therefore, if a low-temperature environment is needed, the payloads can be mounted on the cooling surface of the satellite and the payloads will be cooled by radiation. If the payload requires a high-temperature environment, electrical heating is always the option. If the individual payload unit, like other electronic units on the satellite, does not require special temperature control, the conventional temperature environment that the satellite can provide is generally −20–45 °C. But this does not include temperature control of individual components inside the unit. Therefore, when designing the science payload unit, attention should be paid to the thermal design of its internal circuit boards to avoid making the temperature of the core components exceed the required temperature for normal use.

In order to avoid potential problems in the whole spacecraft test after payloads mounting, the payload unit is usually required to undergo the temperature environment experiments in its own name, including temperature cycling experiments and thermal vacuum experiments.

7.3.3   Power Usage Requirements

The power supply for the science payload comes from the main power supply of the satellite platform. According to technical specifications, the primary power supply of the satellite's power system is +27 V. Power requirements for other voltages can be met by the payload's own power converter or by the power manager of the payload system. In addition, secondary power can be requested from the satellite platform. In addition to the voltage requirements, there are also power requirements of the science payloads that should be put forward during the design phase, and this has become one of the bases to determine the total power of solar panels of the satellite. In the spacecraft design phase, it is necessary to obtain the data of the different power demands of the payloads, such as the variation of the power when the operating mode changes. So there is a difference between the peak power demand and the average power demand. In addition, it is necessary to ensure the magnitude of the current pulse variation, when the payloads are powered on. What needs particular attention is to avoid the so-called current "wave" when several payloads are powered on at the same time. Even if there is only one payload powered on, the current "wave" upon power-on should be controlled, e.g., by connecting a "wave" suppression circuit in series with the power input. Figure 7.13 shows the wave curve after powering on the on-board electronics.

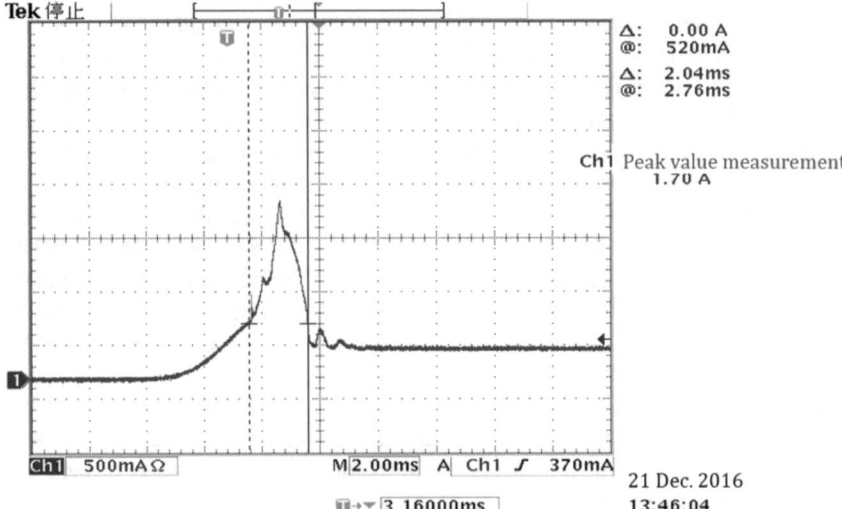

Fig. 7.13   The wave curve after powering on the on-board electronics

7.3.4 Electromagnetic Compatibility Environment Requirements

A mature satellite platform is characterized by sound electromagnetic compatibility (EMC). Clock frequencies, local oscillation frequency of radio frequency (RF) transmitter and receiver, and various harmonic frequencies should be prevented to overlap with each other so as to minimize interference. The on-board cables need to be shielded and the coupling between cables also should be treated. Therefore, in the implementation of the science mission, the satellite platform will impose conventional EMC requirements on the payload [2] to reduce unnecessary electromagnetic radiation and avoid interference with other equipment on-board, as well as the inference between payload units. Before the payload unit is mounted on the satellite, electromagnetic compatibility (EMC) experiments are required to check whether it meets the requirements of EMC use. Figure 7.14 shows the lab of electromagnetic compatibility experiment.

For missions with the scientific objective of detecting electric field, magnetic field, and low-frequency electromagnetic waves in space, the science payloads will pose special requirements for the satellite platform. Besides the requirement of extending the probe out of the satellite body, there are other requirements for special design in terms of magnetic cleanliness and electrical cleanliness. For example, the use of ferromagnetic materials should be avoided in order to reduce the residual magnetic field in the satellite body. In the design phase, a proposal should be made to avoid using ferromagnetic materials for the instrument box and even screws and to replace them with titanium alloys if necessary. The residual magnetism of the entire satellite is measured during satellite assembly. In order to counteract the residual magnetism,

Fig. 7.14 The lab of electromagnetic compatibility experiment

permanent magnets are applied. In order to reduce the surface charge that will create an electrical potential difference, the entire surface of the satellite (including the surface of the solar panels) should be coated with a conductive film to guide the flow of the charges induced during the flight through the space region with high electron density so that the induced charges will not concentrate in certain areas of the satellite surface, and the entire surface of the satellite should be made as equipotential as possible.

7.3.5   Control and Information Usage Requirements

The on-board transponders (receivers and transmitters) are dedicated to transmitting uplink and downlink commands and engineering parameters. Requirements should be proposed for the scientific payloads in terms of command transmission, as well as engineering parameter acquisition and transmission. These requirements are the design inputs for the central computer and the TT&C system. In addition, more often than not, the data volume of scientific exploration is very large, and in case the satellite platform fails to meet the downlink demand to the ground station, it is necessary to equip a data storage device and digital transmitter for the payloads. Necessary coordination with the satellite platform should be made in terms of the mounting position of the digital transmission antenna.

7.3.6   Radiation Environment Requirements

Space science satellites may use various orbits besides the Near Earth Orbit (NEO), and their orbital inclination is not always about 90° in polar orbit, which makes the radiation environment they may go through during their lifetime very different from that of application satellites. The satellite platform will define the total radiation dose to be tolerated by the components. The damage caused by high-energy particle radiation not only occurs during solar eruption events but also on the occasion of flying through the radiation belt, which leads to a single event upset (SEU). If the satellite orbit cuts through the radiation belts many times, the radiation dose on the satellite will be much higher than that of the satellites flying on the Near Earth Orbit (NEO). Therefore, the design of shield from the radiation environment becomes very important. There are two ways to protect against radiation: either by reinforcement, i.e., adding tantalum metal shields to the surface of the components or increasing the thickness of the instrument box panels; or by selecting components that are radiation-resistant, i.e., using highly reliable components for space flight. For the polar orbit satellites, the radiation dose during its lifetime can often reach tens of thousands of krad (Si). If the radiation resistance of the components that to be used in the satellite is yet to know, a simulation experiment of the radiation resistance can be performed on the ground using an accelerator. For components that do not meet the requirements, it is necessary to replace them or add extra protection.

References

1. Yiyong LI, Qiongling SHAO, Xiaojiang LI (2013) Spacecraft payload. National Defense Industry Press, Beijing
2. Fortescue P, Swinerd G, Stark J et al (2011) Spacecraft systems engineering, 4th edn. Wiley, New York

Chapter 8
Technical Fundamentals (III): Mission Planning and Operations

8.1 Introduction

During the development phase of a space science mission, the development of science payloads is the most important part, while during the mission operation phase, the scientific operation takes the leading role. This chapter will focus on the planning and operations of science missions.

In order to better implement space science missions, the ground application system, as one of the five major systems of space systems engineering, are further subdivided into science application system and ground support system. In general, the ground support system undertakes most of the public support work in the ground application system, while the science application system primarily undertakes the technical support associated with a particular science mission.

For a space science mission, it is compulsory to make detailed planning for mission operation at the outset of the mission proposal. The planning is not only useful for the preparation of the operation work after launch but also conducive to avoiding the deviations from the mission and payloads design phase, which may incur the risk of unnecessary omissions, etc. After the launch of a scientific satellite, both the aforementioned two systems need to make short-term, medium-term, and long-term operation plans to carry out the planning formed previously and take care of the data management concerning the detection, observation, and experiment.

8.2 The Application System of Space Science Mission

8.2.1 Six Systems of the Space Science Mission

The space science satellite mission consists of six major systems, namely the satellite (spacecraft) system, the launch vehicle system, the launch site system, the TT&C system, the science application system, and the ground support system, as shown

© Science Press 2021
J. Wu, *Introduction to Space Science*, Springer Aerospace Technology,
https://doi.org/10.1007/978-981-16-5751-1_8

Fig. 8.1 Six major systems
of a space science satellite
mission

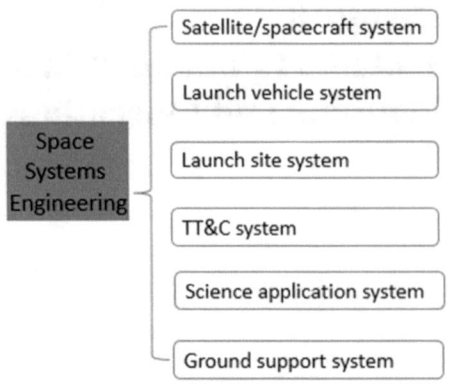

in Fig. 8.1. The first four systems have been touched upon in the previous chapters, so this chapter will mainly deal with the science application system and the ground support system.

8.2.2   Science Application System

The science application system represents the user of the scientific satellites and is responsible for making the science operation plans, including short-term, medium-term, and long-term operation plans, and submitting these plans to the ground support system. The science application system is also responsible for receiving the scientific data and the supporting engineering data from the ground support system, and analyzing and handling the data to form data products at all levels. Although the science application system is responsible for its engineering development and is not directly involved with scientific research and paper publications, it is the direct interface between the principal investigator (PI) and the mission. Therefore, the science application system is also responsible for the development of various data analysis software as proposed by the PI.

The science applications system typically consists of the following subsystems: scientific planning and operation, scientific data handling and storage, and scientific product release and mission results management. The planning and operation and issues related to scientific data handling will be described in detail in this chapter.

The science application system will be built revolving around the PI's home institute, where the mission proposal comes from. But due to the genuine nature of the scientific satellite mission, the science application system will not be located in one institute. The hardware of the system is mainly computers and monitors. It is not necessary to purchase too many special equipments, and caution should be made to avoid duplications in the development of the system.

8.2.3 The Ground Support System

The ground support system is responsible for all the general supporting ground technologies excluding those of the science application system. This system is equivalent to the ground application system of the application satellite mission, and its functions include receiving scientific data via ground stations, pre-handling of scientific data, distributing data to science users, processing mission plans and demands from the science application system, and forming engineering commands for uplink, and other ground support work that all scientific satellites may get involved with, such as contacting data receiving stations, national satellite control centers, synchronization of the timing among the science application systems, as well as the construction of the communication network.

The ground support system is planned and established by the managing institution of the scientific satellite missions and will be used for all scientific satellite missions.

The ground support system usually consists of the following subsystems, including data reception, integrated operation and control, space science data, and communication and network.

8.2.4 System Development Procedure

Once a mission is approved, the science application system and ground support system, whose major task revolves around the software development and configuration of ground hardware, will undergo a different development procedure compared with the satellite (spacecraft) system, as shown in Fig. 8.2.

After the mission approval, for the two systems mentioned above, the mission requirements analysis comes first, which is followed by software and hardware design. As for the software, its development should follow the engineering requirements, and the development phases include requirements analysis, coding, testing,

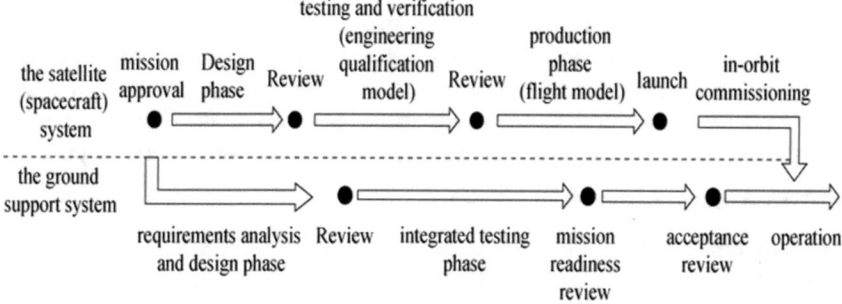

Fig. 8.2 The development timeline between the ground support system and the satellite (spacecraft) system

integration testing, review, code modification, regression testing, etc. As for the hardware, it relies mainly on procurement. In general, the price of the computer decreases over time. In fact, at the same price, the later the procurement takes place, the higher the performance of the hardware. Therefore, hardware is generally procured before testing begins in the later stage of development.

In the final stage of development, two systems will participate in the joint commissioning of all the systems, after which the mission readiness review will take place in parallel with the satellite/spacecraft factory acceptance. After these reviews and acceptance are passed, the satellite/spacecraft will wait for the launch window.

8.3   Planning of Space Science Missions

When proposing a mission, systematic considerations should be given to its future operations from the perspective of spacecraft systems engineering. For example, for the astronomical missions to survey the sky, necessary planning should be made in terms of the sky survey coverage and frequencies of scanning; while for missions that will conduct fixed-point observation of a particular group of objects, considerations should be made in terms of the observation time allocation on each specific target, as well as solar radiation and other stray light interference. It is also necessary to give specific requirements on the attitude and other resources of the spacecraft and consider the required total observation time and spacecraft lifetime constraints, etc. These requirements will influence the overall mission design, so a corresponding study and assessment of these requirements should be carried out as early as possible.

8.3.1   Analysis of the Requirements for Detection and Experiment

At the science mission proposal phase, it is necessary to put forward the requirements for future detection and experiments that facilitate the materialization of the scientific objectives. The more clearly the requirements are defined and formulated, the clearer the design inputs for the spacecraft and other systems. Clear design inputs are essential not only for the satellite and payloads hardware development but also for the realization of the scientific objectives during the mission operation phase.

(1) Astronomical Observation Missions

For astronomical observation missions, it is required to specify the telescope's observation direction, which is usually divided into survey and selected target observation.

(I) Survey

A survey means that the telescope's field of view should cover the entire celestial sphere (or part of it) in a time period as short as possible. For satellites with a three-axis stabilization attitude flying in the Earth orbit, as the satellite itself rotates around the Earth, the telescope pointing to the zenith provides a 360° strip of coverage of the celestial sphere in each circle, and the width of the strip is the field of view of the telescope. If the field of view is adjusted by pointing the telescope perpendicular to the orbital direction, another strip of coverage can be achieved in the next circle. And then, the two strips can be stitched together. Of course, the most efficient survey with the consumption of the least amount of resources can be realized by the optimization of orbit and attitude, if the precession of orbit in inertial space and the rotation period of the Earth around the Sun is taken into consideration. It is self-evident that the Earth orbit is more suitable for survey missions. For the survey mission that flies in the interplanetary orbits, the slow spin strategy should be adopted by the spacecraft to achieve efficient surveys. Figure 8.3 shows the survey diagram of NASA's Transiting Exoplanets Survey Satellite (TESS) mission.

(II) Selected Target Observation

Selected target observation refers to the continuous observation over a long period of time with the telescope pointing in one or several specific directions, such as the central region of the Milky Way. For a spacecraft flying in inertial space, without

Fig. 8.3 The survey diagram of NASA's TESS mission

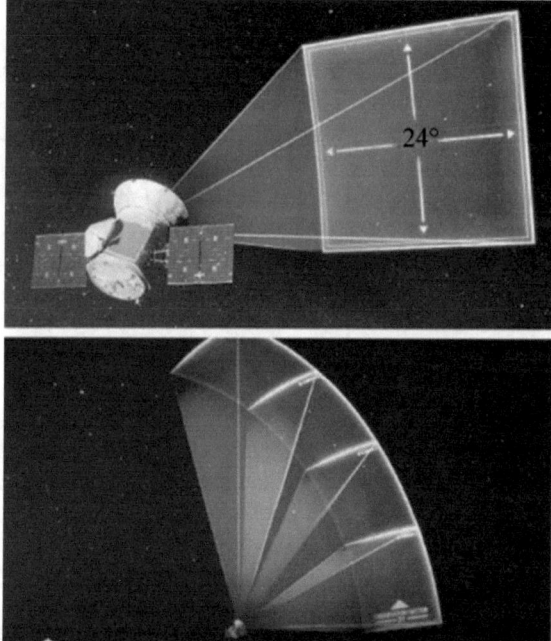

perturbations, its attitude inherently leads to the pointing in a particular spatial direction. Therefore, no matter what kind of orbit a spacecraft may fly in, selected target observation is the easiest to realize. Selected target observation usually requires continuous observation in a particular direction for a certain period of time, and then turning to the next selected target, which requires necessary attitude adjustments. Bear in mind that such adjustments are not finished in a glimpse. The adjustment process takes time, and it also takes time for the attitude to stabilize after the necessary braking maneuver, which must be taken into consideration during the mission planning. Therefore, if the number of targets to be observed is large, it is necessary to optimize the observation sequence to avoid unnecessary time spent on attitude adjustment and stabilization, as well as the unnecessary consumption of fuels.

(2) Space Exploration Missions

The space exploration mission refers to the in situ or remote sensing measurement of the space electromagnetic field and particle environment.

(I) In situ Measurement

In situ measurement is conducted by spacecraft through reaching particular locations or regions in space. Therefore, the planning of the exploration mission is closely related to the orbit design, which must allow the spacecraft to traverse the targeted area to be detected. The probes should be switched on while traversing the area so as to perform the measurements. In addition, planning should be made to differentiate the data level of importance, such as when high-resolution data is needed and when low-resolution data is sufficient. For a multi-point detection mission, the planning should focus on how to maximize the time of flybys of different spacecraft and the time of coordinated joint detection. Therefore, geospace exploration missions often adopt elliptical orbits. The spacecraft's flight velocity is the slowest at the apogee of the elliptical orbit, and the longest observation time can be obtained if the apogee of the orbit is designed to be within the observation region. For in situ measurement, sometimes the measurement to all directions around the spacecraft is needed, which explains why these missions often adopt a simple spin-stabilized attitude in order to obtain the observation in 360°. If a probe has to be mounted on a spacecraft with a three-axis stabilization attitude, it is necessary to mount the probe in the direction of particle incidence.

(II) Remote Sensing

Remote sensing detection refers to the remote measurement of the electromagnetic field and particle parameters of the space environment by detectors with long distances, such as neutral atom imagers and extreme ultraviolet imagers. The observation requirements are similar to those of astronomical fixed-point observations missions. In case the environment surrounding the observation region is not of interest to the mission, it is necessary to plan ahead, i.e., to shut down the observation window when it is not facing the region. Remote sensing observation requires intensive planning in terms of satellite attitudes and it has very high requirements for orbital positions.

(III) Earth, Planetary, and Small-Body Exploration Missions

This type of mission often orients the probe toward the target object to be observed, and its major task is to make global imaging of the targeted celestial body or to survey a specific region in detail. For example, the main objective of Chang'e-1 lunar mission is to obtain a global moon map, while Chang'e-2 is to survey in detail with high resolution the future landing area of Chang'e-3. The main consideration for this type of mission is solar illumination.

For microwave remote sensing detection [1], there is no need to consider daylight exposure. However, for active microwave remote sensor that has transmitting function, such as scatterometer and synthetic aperture radar, the requirement for transmitting power should be considered.

(IV) Microgravity Science, Life Sciences, and Space Fundamental Physics Experiments

The major requirement of this type of mission is the microgravity environment. Any mechanical movement and vibration of the flying spacecraft, even the vibration of the solar panel, will cause a decrease in the microgravity level. Therefore, it is necessary to select the best timing for this kind of science experiment to avoid any spacecraft maneuvers and other mechanical vibrations. In addition, attention should be paid to switch on/off operations and power consumption allocation, because a large number of materials science experiments require high-powered high-temperature furnaces. For example, a crystal growth furnace probably requires several kilowatts of power, and once it is powered on, almost all of the power on the satellite has to be supplied to the crystal growth furnace. If it is a dedicated microgravity scientific experiment satellite mission, solar panels are always avoided, which is replaced by a large battery determined by the power requirements and experiment time so as to make the satellite microgravity level meet the requirements of the on-board experiments.

8.3.2 The Spacecraft Conditions and Resource Constraints

Although the design of space science missions focuses on scientific objectives, not all scientific mission requirements will be satisfied without conditions. This is due to the spacecraft's own technical rules and limitations of available resources. Therefore, obtaining the requirements of science mission operations as early as possible and using them as design input is the basis for designers and scientists to get the best design results through repeated discussions and design updates.

For scientists, while presenting clear scientific operation requirements, their needs will be more reasonable if they understand the technical constraints and resource limitations of the spacecraft design.

The following technical and resource constraints are critical:

(1) Orbital Dynamics Constraints

Once the spacecraft is launched into orbit, it is unrealistic to change the orbit. This is because the vehicle travels very fast with a huge amount of kinetic energy, and changing its direction of motion will consume substantial energy at the cost of a large amount of fuel or electricity. In addition, two spacecraft cannot fly in parallel in Near Earth Orbit, because the center of the orbit must be at the center of mass of the Earth, leading to the eventual orbit intersection of the two spacecraft at a certain point. However, it is easier to achieve spacecraft formation flying in the same orbit. This should be taken into account when designing the spacecraft formation.

(2) Attitude Dynamics Constraints

Changing attitude is much easier than changing orbit. However, since there is no air drag in space, any perturbation to an established attitude requires the same amount of energy to re-stabilize the attitude. So bear in mind that although the energy required to change the attitude is not that much, twice as much energy is required to stabilize it. What's more, it takes time to change the attitude in that stabilizing the attitude takes much longer time than just changing the attitude. This technical constraint should be considered especially for missions requiring high-precision attitude stabilization.

(3) Energy Constraints

In orbital flight, the spacecraft mainly relies on solar energy. The size of the solar panel to be designed depends on the energy requirements of the mission. Larger solar panels can provide more energy, but at the same time also increase the launch cost due to the mass increase. In addition, for solar system probes, it is necessary to consider a couple of factors, such as the distance between the spacecraft and the Sun, the angle of solar incidence, etc. The closer the spacecraft to the Sun, the more power the solar panel will produce and vice versa. In order to increase the power output of the solar panel, it is necessary to keep the angle of the direction of solar incidence and the solar panel's normal direction as constant as possible within a small range. Therefore, if a science mission requires a large amount of energy, this energy requirement should be iterated clearly at the mission design phase, which will be the input for the mission designer to optimize the energy allocation after holistic considerations. Science missions featuring microgravity materials science typically require a large amount of energy, such as heating of materials. For high-power microwave radar, it is necessary to consider the power constraints. For solar system probes to the outer planets, there is a need to increase the size of the solar panel or even use nuclear energy, because the solar energy obtained per square centimeter is decreasing. As for the moon landing mission, it may experience a long lunar night (14 days) on the lunar surface, during which no solar energy is available. In this scenario, nuclear energy is often the complementary option to ensure the minimum need for the spacecraft/probe's survival. In times of high energy demand, the supply more often resorts to the practice of allocating time slots to specific scientific tasks or even shut down altogether.

(4) Limitations on the Amount of Data that can be Downlinked

The science output of a science mission rests on its scientific data, making it a valuable resource. Therefore, in general, valuable scientific data should be downlinked to the maximum extent possible. However, the time for a flying spacecraft in orbit passing the ground station is limited, and the farther away from the Earth, the weaker the signal will be, leading to less transmitted data. For a spacecraft orbiting in Near Earth Orbit, the pass time to a ground station is a few minutes up to 20 min, during which the amount of data that can be downlinked is very limited. The ways to increase the amount of downlinked data include increasing the antenna aperture, increasing the transmission frequency, changing the coding mode to increase the bandwidth, compressing the data, and adding ground stations in different geographic locations to increase the pass opportunity and time. When these technical methods reach the limits of available resources, it is compulsory to consider from the perspective of mission planning how to reduce the requirements for downlinked data. The normal practice is to perform on-board data handling (OBDH) to first discard data that is useless or less scientifically significant, and then compress the remaining data. Even after this, if the data set is still too big to transmit, all data must be classified according to the importance and then selectively downlinked. The impact of the data loss on the materialization of scientific objectives must be reviewed at the early stage of mission design.

(5) Spatial and Temporal Reference Constraints

High-precision fundamental physics experiments, such as gravitational wave detection and space VLBI observations, require high-precision spatial and temporal references. Therefore, it is necessary to consider whether the same experiment conditions can be obtained in space as on the ground. Due to resource constraints, the conditions on spacecraft that can be achieved with current technology are not comparable to those on the ground, which must be taken into account at the time of the mission proposal. In terms of time reference, small atomic clocks can be installed on the spacecraft to provide a time standard that can be updated and calibrated by the time service from the ground. The time accuracy of on-board crystal oscillators is usually 10^{-9} s/day, but the use of an atomic clock can greatly improve the accuracy. For instance, the time accuracy of the atomic clock on-board China's Tiangong-2 space laboratory has reached 10^{-15} s/day.

8.3.3 Compiling and Execution of Mission Plans

As mentioned earlier, at the initial stage of space science mission design, a clear mission planning input is the basis for the success of the entire mission. Once the spacecraft is launched into orbit and the follow-up in-orbit commissioning is completed, it enters the mission operation phase. During the operation phase, the operation plan is compiled in accordance with the mission requirements. The operation plan is the basis for the execution of scientific observation, detection, and experiments.

The main contents of the mission planning include following the time order of spacecraft's position in orbit, spacecraft attitude adjustment commands, science instrument switch-on/off commands, observation mode adjustment commands, science data on-demand (i.e., determine time period to transmit science data), and engineering parameter data (housekeeping data) transmission commands, etc.

Conventional mission planning comprises long-term, medium-term, and short-term planning. Long-term planning is based on the scientific objectives and the mission requirements upon the mission approval, and it is usually divided into annual plans. Long-term mission planning is subject to adjustments after the mission execution for a period of time, which takes place, for instance, when certain tasks are completed earlier or when certain tasks need additional time. Medium-term mission planning, a more accurate plan, is usually made on a quarterly to half-year basis. Based on the medium-term mission planning, the ground support system will have a general arrangement of the workflow for the operation staff. Short-term planning, done on a weekly basis, is specific observation, detection, or experiment requirements, which will be provided by the science application system to the ground support system in charge of translating the requirements into uplink commands to be sent to the satellite control center for uplink. In the absence of contingency, such as malfunction in an instrument, short-term planning is not subject to adjustment. It is always formulated in advance on a weekly basis.

According to the labor division in mission planning in a space science mission, the science team led by the principal investigator (PI) proposes the science requirements, which are documented by the science application system and transmitted to the Science Mission Operation Center of the ground support system. The center then checks the engineering feasibility. If no problems are found, the software will translate the planning into uplink commands and transmit it to the satellite control center of the TT&C system for uplink to the spacecraft for execution; if problems are found, communication will be done with the science application system for verification and necessary correction.

China Xi'an Satellite Control Center is responsible for sending uplink commands and receiving downlink engineering parameters from the satellite, and later transmitting them to the ground support system. For China's manned space program and China's lunar exploration program, this function is performed through the Beijing Aerospace Control Center (BACC).

8.4 Science Data Reception

Scientific data are at the heart of a space science mission. As the spacecraft passes the ground stations that are responsible for receiving the science data, the science data will be downlinked according to the uploaded mission planning. Once received by the ground stations, the data is backed up on-site, and at the same time, the data is transmitted via the ground communication network to the ground support system, the Science Data Center in Beijing, for pre-processing/handling, where the Level 0 data will be updated to Level 1 data. In this process, the supporting engineering parameters transmitted from the satellite control center will be complemented, such as the orbit parameters, time, and attitude. In general, there is no requirement for real-time production of Level 1 data, but the production should be done as soon as possible and the data product should be sent to the science application system in time.

8.4.1 Ground Station for Science Data Reception

The ground station for science data reception has a dedicated antenna to receive data from scientific satellite missions. The aperture of the ground station antenna is difficult to change once its design is frozen, and in fact, one antenna is usually used for multiple scientific missions. There are three main ground stations in China that can be used for data reception for scientific satellite missions, namely Miyun Station, Sanya Station, and Kashgar Station. Figure 8.4 shows the antenna of China's ground station for science data reception in Sanya, Hainan province.

The antenna transmits the received signals to the microwave receiver of the ground station, amplifies, and demodulates them into baseband signals, which are then

Fig. 8.4 The antenna of China's ground station in Sanya

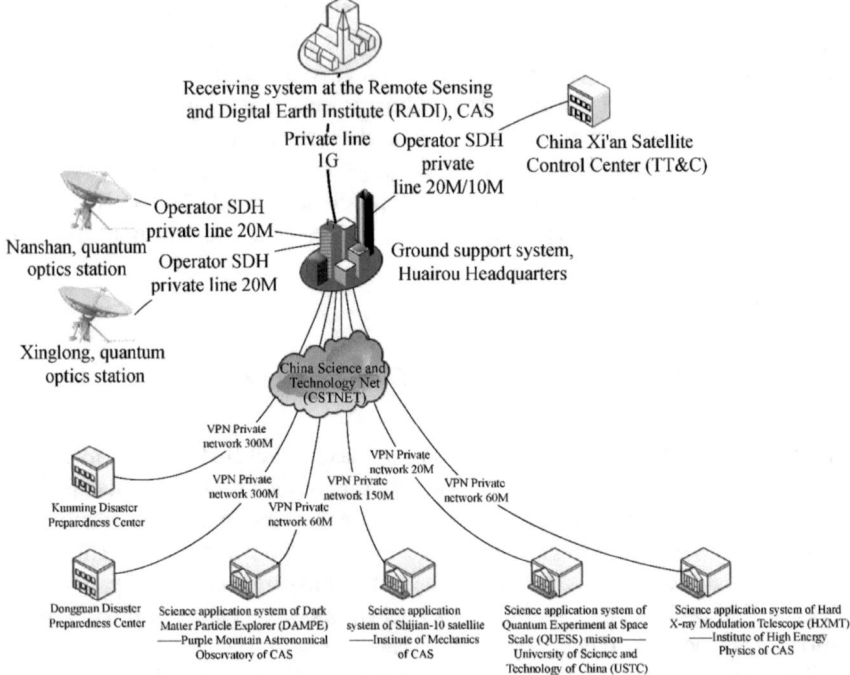

Fig. 8.5 Science data transmitted in real time to the ground support system in Beijing

digitized, stored locally, and sent to the science data center at the ground support system headquarters through the ground communication network. Figure 8.5 shows the ground communication network of the ground support system of the Strategic Priority Program on Space Science (Phase I) of the Chinese Academy of Sciences.

8.4.2 Spacecraft Pass Time and Downlink Rate

The pass time of spacecraft, depending on the orbit design, is a very important part of the data reception process [2]. Since the altitude of Near Earth Orbit is relatively low, the pass time is very short, from a few minutes up to 20 min. For elliptical orbits, there is a longer pass time, up to several hours at apogee and down to a few minutes at perigee. The distance between the apogee and the Earth is really big, and the data transmission rate will decrease despite the long pass time.

As for the constraints of the pass time, it is defined as the time with an elevation angle greater than 5°, and an elevation angle less than 5° would not be counted. The maximum pass time in Near Earth Orbit occurs when the spacecraft is right over the zenith.

The aperture of the receiving antenna is one of the design constraints on the amount of data that can be downlinked from a space science mission. However, with technological advancements, it is possible to increase the frequency used for data downlink without changing the reflector of the receiving antenna. For the time being, the operating frequency of the receiving antenna for scientific satellite missions can be upgraded from S-band to X-band or even Ka-band. Correspondingly, the design of the data transmitter and antenna on scientific satellites needs to be changed.

8.4.3 Scientific Data Pre-handling/processing

The handling/processing of the science data needs the complementary engineering parameters at the time of observation/detection, such as time, orbital position, attitude, etc. These engineering parameters can be packaged and downlinked through the TT&C system together with the science data. The ground support system will process the engineering parameters of the satellite platform (bus), the engineering parameters of the payload, including the operating parameters of the science payloads such as temperature, voltage, current, and the science data in a separate manner, and prepare these parameters in a certain format to produce Level 1 data. This process is called *pre-handling or pre-processing*.

8.5 Science Data Classification and Distribution

8.5.1 Science Data Classification

According to the degree of data processing, science data can be categorized as Level 0, Level 1, Level 2, Level 3, etc.

Level 0 data is the unprocessed data received by the ground station, which will be transmitted in real time to the science data center of the ground support system.

Level 1 data is the science data product that separates the science data from engineering parameters with annotations of time and related engineering parameters (including payload engineering parameters). It is processed by the ground support system and sent to the science application system as soon as it is generated.

Level 2 data is usually basic data concerning individual science payload, which has undergone preliminary calibration for scientific research. Level 2 science data is usually processed and generated jointly by the payloads development institutes and the science application system under the leadership of the principal investigator (PI).

Level 3 data is tabular and pictorial science data that can be easily identified and used in conjunction with spatial or geographic displays. It is usually generated by the science application system under the leadership of the PI. As part of the mission output, science data published in articles are usually Level 3 science data. Even after publication, Level 3 science data can also be used for deeper data mining and exploitation.

For more detailed classification, science data can also be further subdivided into 1A, 1B, 2A, 2B, etc. The definition of data classification can vary from one scientific mission to another, but the data classification plan must be released through formal documents to facilitate the data usage by the users.

8.5.2   Science Data Distribution

As a normal practice, the data is released by the approved institution of a particular satellite mission. According to the approved institution's requirement, Level 1 science data will be first distributed to the institute that is responsible for the mission proposal and the science applications—usually the home institute of the PI where the science application system is located. The science application system is responsible for generating science data products at Level 2 and above. In accordance with the data policy and the principle of maximizing output, science data products should be made available to the scientific research community as soon as possible. The way to open the data varies from mission to mission. For users, the most common way is to register and request access to the data through the mission website. The science application system is responsible for the development of the science data website.

8.5.3   Data Policy

Once the data products are generated, the distribution of these products follows the mission's data policy. In general, there are two data policy options.

The first data policy favors the motivation of the mission team: the access priority and exclusive rights to the data is the reward to the mission team for their contributions over the years from the mission proposal all the way to the mission operations. In general, researchers who proposed the scientific objectives are the most motivated and task-oriented in the usage of data, and they are most likely to make significant scientific discoveries. Furthermore, when the data are not of good quality or not ready for optimal use, the motivation of those outside of the mission team to use the data is compromised, leading to low data utilization and significantly reduced scientific output. Therefore, exclusive rights for a certain period of time not only help protecting the enthusiasm of the mission teams, giving them time to improve both data quality and user-friendly tools for the analysis but also maximizes the amount and quality of the science output of the mission.

The second data policy is full-open data policy, which attaches importance to promoting scientific output by making data available to a wider community. It is more applicable for stable data flows produced by observatory missions in survey mode. This is especially the case for the missions that their observing objects are constantly changing, leading to a steady stream of scientific output or even applications, such as Earth observation missions and solar observation missions. For space science missions, the constant changes occurring on the observing objects will produce a large amount of new data every day, which is beyond the processing capacity of the mission team. By making the data fully open, the potential benefits of the data can be maximized, producing a more substantial scientific output.

In general, for space science missions, the data policy is individually tailored. The usual practice is that the first data policy is adopted for a period of time, such as six months to a year, and then the second policy of open data will be executed depending on how the situation evolves itself. The length of the period of exclusive data rights depends on the specific mission and the nature of the data and follows the requirement of maximizing the scientific output.

8.5.4 Science Data Archiving

All levels of science data products are national property. They are valuable assets for the science team to carry out scientific research and follow-up deeper data mining. Therefore, the archival management related to the data products should be strictly regulated. The ground support system is responsible for the archiving and management of all science data on behalf of the mission approval institution.

References

1. Jianjiang W, HU Xuejun (2019) Uncertain mission planning of earth observation satellite. Science Press, Beijing
2. Uhlig T, Sellmaier F, Schmidhuber M (2015) Spacecraft operations. Springer, Berlin

Chapter 9
Management (I): Call for Mission Proposals and Its Selection

9.1 Introduction

The lifecycle of a space science mission begins with the mission proposal. We know that not all of the missions proposed will get approved and funded, and the number of proposals received is always far larger than the budget available, which is not only the case in space-developed countries/regions like the USA and Europe but also in space developing countries like China. Therefore selection is a necessary process to invest in those missions with better potential for a great science impact and involvement of a motivated community. A limited ratio of successful missions indicates a really competitive selection process, ensures world-class science, and makes the program sustainable [1].

This chapter will focus on the mission proposal and selection phase and presents during the phase what kind of problems a science team, a technology team, and a management team will meet and how they should deal with them.

9.2 Identification of Science Questions

From the governmental agency perspective, on the one hand, to receive the most innovative mission proposals, it shall organize the scientists to conduct the strategic planning that focuses on the scientific questions; on the other hand, the agency shall allocate the necessary and limited funding also through strategic planning.

From the perspective of the science community, the science teams or individuals are usually limited in their own disciplines, leading to different opinions about the frontiers which need to be addressed first in a broad space science framework. They may prefer the science questions in their own discipline to be the most important scientific challenge. Therefore, strategic planning is necessary for the community to reach a full consensus about the priorities, and get enough funding to implement and launch the mission.

© Science Press 2021
J. Wu, *Introduction to Space Science*, Springer Aerospace Technology,
https://doi.org/10.1007/978-981-16-5751-1_9

It is thus clear that, for both the governmental agency and science community, the first thing to do is strategic planning, if they want to have great space science missions to be proposed. Through planning they can see in which fields their peers around the world are working; what are the unsolved but important "bottleneck" questions; what are the scientific questions resulted from the contradiction between newly achieved discoveries and the existing theory; what are the fields with promising applications if the breakthrough is achieved; which scientific questions will be addressed only with space tools; which scientific questions meet the national demands; which science questions are so ambitious that they need to be addressed through step by step observations and gradual technology capacity building. The process of reaching a consensus is the process for both science community and government agency to improve their own capability.

9.2.1   Strategic Planning

Since the beginning of space age in 1957, humans have got a better understanding of the universe, the solar system, especially the Earth we are living on. However, it dawns upon us that more knowledge comes with more questions. Then, which questions out of those are the most important ones? What breakthroughs would fundamentally change the human's understanding of the law of nature? The more accurate our identification of those questions, the more helpful it is to find the starting point to address them.

When it comes to the organization practice of planning, it usually gets involved the representative science teams from all disciplines of space science. The first phase of planning, therefore, is to organize workshops in a bid to identify science frontiers. Besides prestigious leaders, a large number of young scientists shall be engaged in discussions and brainstorming. The output of the work during this phase is a strategic plan and an announcement of opportunity report.

The second phase of planning is to issue a call for mission proposals. The mission proposal only includes the preliminary mission concept and payload complement. Workshops could be organized to discuss the proposals submitted. In the workshops, mission proposers introduce their proposals and exchange ideas with other proposers, to improve their own, or even merge into a new one, usually with more innovative and challenging scientific objectives.

The strategic planning report only stands for the identification of science frontiers for a period, while the development of science frontiers is fast, e.g., there are up to dozens of science satellites in orbit now, and large amounts of science data are being analyzed. Therefore, the identification of science frontiers needs to be updated in due time.

9.2.2 Space Science Planning in the USA

In 1863, the National Academy of Sciences (NAS) was founded under the Act of Congress signed by Abraham Lincoln, to provide independent advice for U.S. science, technology, and medicine. NASA also followed the Act since its establishment in 1958, i.e., NASA should at first consult NAS in the case of large science missions planning. Thereby, NAS sets up the Space Studies Board (SSB) in 1958 which is entrusted by NASA to organize planning activities about studies and discussions on each discipline of space science.

The planning, for each space science discipline, is done approximately once a decade, and the findings are published as *Decadal Survey* [2–5], as shown in Fig. 9.1. To adapt to the rapid development of the disciplines, a mid-term assessment is organized to review and adjust the text in the middle of 10 years.

It takes 2 years for every planning, starting from the initial workshops to the publishing of the reports. The panel members, including the chair, need to be changed every planning. Regarding the procedure, the first step is to call for mission proposals (1–2 pages), the so-called "white paper", to collect scientific ideas in each discipline. Then a series of workshops are organized to compile the strategic planning report. The report serves as the basis for NASA to draw up future space science programs and select future missions. According to incomplete statistics, NASA takes even more than 90% of recommendations listed in the *Decadal Surveys* as the driving input when implementing science missions.

There are advantages in adopting this structure. Firstly, it engages the science community in discussions and follows the bottom-up principle which is conducive to reaching consensus on the basis of science excellence. Secondly, it brings the U.S.'s research excellence in space science into full play and will not lead to the situation that some *prestigious* figures control a certain discipline, because the panel members need to be changed every "decade". Thirdly, the mid-term review in a decade could adapt to the development of science frontiers. However, there are some disadvantages, such as some directions need to be planned more than 10 years in advance, making the 10-year cycle a bit short, while for some directions with fast development, the 10-year cycle is a bit long. What's worth mentioning is that, in spite of mid-term assessments conducted in recent years, it still fails to make substantial modifications to the *Decadal Survey*.

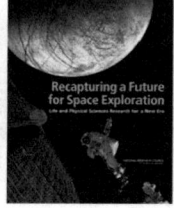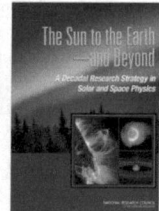

Fig. 9.1 *Decadal Survey* report of space science disciplines

9.2.3   Space Science Planning in Europe

Space science missions from most of the European countries are executed by ESA
as the agency. ESA uses its Science Program Committee (SPC) and Space Science
Advisory Committee (SSAC) to plan its missions, and the former is the decision-
making body, while the latter, with subcommittees for different disciplines, provides
advice and organizes workshops. ESA's space science missions mainly fall into three
categories, i.e. Large (L-Class), previously called corner-stones or ESA-led flagship
missions, Medium (M-Class), and Small (S-Class) providing flexibility to respond
to new challenges. The missions are, respectively, selected and approved every 4–
5 years, 3, and 2 years. Sometimes there are overlapping in the three categories. Every
1–2 years, the announcement of opportunities is issued to call for mission proposals
in different categories. Generally, dozens of proposals will face intensive selection
competition. As to the funding, M-Class mission is about 500 million euros for every
mission, and L-Class 1 billion euros. L-Class missions are dominated by ESA and
open to international partners. Since the founding of ESA, two medium- and long-
term space science plans have been compiled, i.e., *Horizon 2000*, completed in the
1980s and adjusted in the 1990s, and *Cosmic Vision* (shown in Fig. 9.2) [6] published
in 2005 on the basis of *Horizon 2000*. In*Cosmic Vision* the L-Class missions extend
beyond 2030. Currently, a new round of space science mission planning is unfolding
to define the long-term planning for a period up to 2050, thus called *Voyage 2050*.

   ESA's space science missions are also proposed following the bottom-up principle. L-Class missions aim at the future development, M-Class the medium-term
development, and S-Class the flexible and fast development. We can see that the

Fig. 9.2 The ESA *Cosmic Vision* in English and Chinese

structure is adaptive to some extent. However, on the other hand, for ESA's science missions, the mission adoption happens only when all the member state representatives agree unanimously since science missions are mandatory and their funding comes from the ESA's member states. In view of the fact that outstanding and innovative proposals usually come from smaller mission teams, under this structure, the proposals from smaller teams or countries usually cannot get support from all the member states, while missions that get support from all the member states are those with more nations involved and with bigger scientific objectives. Therefore, the missions approved are those moderate ones with compromised consideration. In fact, the statistics about the Nobel Prize winners whose research is based on space science missions' data indicate that most of these missions come from the USA, more than a dozen in total, while ESA's space science missions haven't produced any Nobel Prize winner.

9.2.4 Space Science Planning in China

The Chinese Academy of Sciences (CAS) has been leading the space science strategic planning since 2006. Following the bottom-up principle, CAS has released call for missions and organized strategic planning workshops, which led to the publication and release of some strategic planning reports, including *Space Science & Technology in China: A Roadmap to 2050*, *the Medium and Long Term Plan of China's Space Science Missions (2020–2025)*, and *Calling Taikong—A Study Report on the Future Space Science Program in China* (as shown in Fig. 9.3). CAS organizes space science planning every 2–3 years to call for mission proposals. What's worth mentioning is that the missions being implemented under the framework of CAS Strategic Priority Program (SPP) on Space Science have their origin from the above-mentioned strategic planning. Chapter 14 will introduce in detail the planned 23 space science programs in China.

Fig. 9.3 The study reports on space science programs in China

9.3 Study of Scientific Objectives

An appealing space science mission that is successfully selected and funded must have clear and outstanding scientific objectives.

Both the management team and science team should pay enough attention to the scientific objectives, which is the drive for sustainable development of space science programs.

For the management team, it is necessary to lead the science team to propose scientific objectives with great scientific impact. For the science team, they should propose the objectives based on the scientific questions to be addressed, which is the prerequisite of future science output.

9.3.1 How to Propose Scientific Objectives

To propose clear scientific objectives with great impact, it is necessary to start with science questions. A couple of "why" needs to be asked, which will lead to the scientific objectives of a mission. The realization of these scientific objectives will answer the key science questions. The science questions, as herein mentioned, are yielded through the consensus reached during the strategic planning phase.

9.3.2 The Realizability of Scientific Objectives

On the one hand, scientific objectives would appear to be too specific if they come directly from science questions. In this case, they are easy to be answered but will produce a minor scientific impact. On the other hand, the realizability of scientific objectives would be at stake if an ambitious objective is directly proposed without discussing the specific science questions to be addressed, in spite that the ambitious objective is easier to get the spotlight. Therefore, the process of studying the scientific objectives is to study its impact and realizability.

9.3.3 The Impact

The impact of scientific objectives is manifested in that the mission aims at major science challenges, and the potential breakthroughs can fundamentally change the human's understanding of nature, e.g., addressing the origin and evolution of the universe, dark matter, dark energy, gravitational wave, search for exoplanets, the mechanism of solar eruption, and the trend of global change. Sometimes, wording like "Nobel class" is used to describe the impact of scientific objectives. The impact of science output is also evaluated with the papers published in world-renowned journals.

9.3.4 The Involvement

The other requirement for scientific objectives is involvement, which is as important as the impact. Involvement means the mission's contribution to a discipline's development. How many people are involved in data analysis after the launch of a mission? How many science papers are expected to be produced? To be brief, more involvement, more science output. If we focus only on the impact, we may neglect the fact that the development of disciplines relies not only on a couple of great breakthroughs but also on plenty of follow-up theory improvements after the breakthroughs are made, which involves more people. To put it in a figurative way, after the front of the battlefield was broken by the elite force, a complete victory will be secured only when the following troop clears the whole battlefield. If we say the impact represents quality, then involvement represents the quantity. Therefore, the quantity of science papers is usually used to evaluate the involvement.

9.4 Selection of Payloads

It may seem logical that the study of scientific objectives happens before the selection of payloads. However, in fact, they proceed simultaneously. The reason is that to realize scientific objectives with great impact, the mission has to work out an innovative mission design with better observation and detection ability, which mainly relies on science payloads.

For a mission with a single payload, the payload design has to be innovative so as to provide new observation, detection, and experiment data. For a mission with multiple payloads, if the payloads are not innovative individually, then the mission team need to work out different combination of the payloads, to get innovative new data. For example, Yinghuo-1, the Chinese Mars probe, combined three payloads together, i.e., magnetometer, high-energy electron detector, and particle detector, to identify the Mars particle's precipitation and escape mechanism. It is a pity that the departure engine of Phobos-Grunt, the spacecraft on which Yonghuo-1 was piggybacked, was not ignited, and Yinhuo-1 failed to go to Mars. In the case of the innovation of an individual payload, technology breakthroughs are needed. For example, when adopting the "occultation" method to search for exoplanets, a photometer with high sensitivity was necessary, which requires the payload team to achieve technology breakthroughs, such as improving the sensitivity from 1/1,000 to 1/10,000.

9.5   Mission Profile

The profile of a mission includes the orbit requirement, the observation/detection/experiment scheme, the required environmental conditions, the data storage and downlink, and whether the ground experiments are required, etc. For example, when the astrometry method is used to observe the movement of the host star, a catalog is needed to select the host star and several reference stars to determine the aperture of the telescope to observe whether the host star owns planets, the number of planets, the movement period, the distance of planets from the host star, the measurement of time, etc. The clearer the description, the more helpful it is for the mission selection and future engineering design.

9.6   Payloads' Requirement for the Spacecraft

The science team needs to quantify the payloads' requirements for the spacecraft, including the mission plan, the mass of payloads, power consumption, attitude control, installment and mounting requirements, and data downlink capacity.

To compile the requirement report, cooperation between the science team and the engineering team is needed. The clearer the requirements proposed by scientists, the more helpful they are for the engineering team to take them into consideration earlier. The engineering development phase, especially in the early stage, is a process during which the science team and the engineering team closely work together to make constant adjustments. The engineers need to understand the scientific objectives and the working principle of science payloads, while the science team needs to understand the technical constraints and feasibility. When the existing technology cannot meet the requirements of the science, the parties need to discuss it over and over. Both the attitude of ignoring feasibility to insist on high sensitivity and the attitude of reluctance for innovation should be avoided.

9.7   Selection of Mission Proposals

The above-mentioned points in this chapter are the cornerstone of a science mission proposal, with scientific objectives as the most important part, which is seconded by the engineering feasibility.

The management team needs to organize the selection to select the best ones out of the received proposals. On the other hand, the science team needs to make their own proposals excellent enough to get selected.

The selection process must be open, fair, and unbiased. To make it happen, the agency needs to release the selection criteria beforehand to all the mission teams, and organize selection meetings. The selection panel should not include the mission team members. The selection process is shown in Fig. 9.4.

For the space science missions selection, domestic and international missions alike, the main criteria are (1) the impact and ambitions of the scientific objectives, i.e., whether a mission aims at major science challenges, and whether the potential breakthroughs can fundamentally change human's understanding of natural laws; (2) the involvement of excellent science teams in achieving those goals, i.e., whether a mission is supported by a significant number of high-level researchers who are involved in the data analysis and use the scientific observation and experiment capacity of the platform to produce large amounts of good science. The former focuses on whether it can achieve great scientific breakthroughs, while the latter focuses on the contribution to the discipline's development and the involvement of more scientists in data analysis [7].

A mission that meets both criteria has better odds to get selected. The Hubble Space Telescope (HST) acts as a typical example. It was an optical telescope with the largest aperture back then and with scientific objectives of great impact, aiming to address the origin and evolution of the universe. With HST, scientists have achieved great scientific findings, such as verification of the origin of the Big Bang, and validation of the accelerating universe to confirm the existence of dark energy. Meanwhile, HST is an observatory that could serve a large number of research teams, with great involvement of the community. As a matter of fact, it brings about the growth of a generation of space astronomy teams. Since the launch, the number of high-level papers published using its data has exceeded 20,000. The James Webb Space Telescope (JWST), the successor of HST, is planned to be launched in 2021. It is also a mission with both great impact and involvement. In the latest NASA *Decadal Survey* in astronomy, JWST has the highest priority.

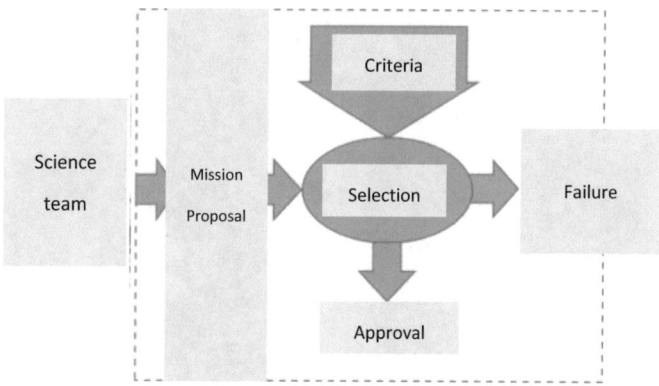

Fig. 9.4 The selection process of mission proposals

References

1. Ji WU (2018) Characteristics and managements of space science missions. Chin J Space Sci 38(2):139–146
2. National Research Council of the National Academies (2010) New worlds, new horizons in astronomy and astrophysics. The National Academies Press, Washington
3. National Research Council of the National Academies (2011) Vision and voyages for planetary science in the decade 2013–2022. The National Academies Press, Washington
4. National Research Council of the National Academies (2013) Solar and space physics: a science for a technological society. The National Academies Press, Washington
5. National Research Council of the National Academies (2018) Thriving on our changing planet: a decadal strategy for earth observation from space. The National Academies Press, Washington
6. European Space Agency (2005) Cosmic vision: space science for Europe 2015–2025. ESA Publications Division, Noordwijk
7. Wu J, Giménez A (2020) On the maximization of the science output of space missions. Space Sci Rev. https://doi.org/10.1007/S11214-019-0628-4

Chapter 10
Management (II): Mission Development and the Duty of Scientists and Engineers

10.1 Introduction

Once a space science project is officially approved, it enters the engineering development phase [1]. The development process of a space science project is basically the same as other aerospace projects, but with its own characteristics, especially that the science team needs to be entitled to play a necessary role in the engineering development so as to deliver the best science output. This chapter will focus on the mission development process of a space science project and the duty of scientists and engineers in the process.

Figure 10.1 shows the full mission lifecycle of a space science mission, covering all the phases that start with strategic planning. The previous chapter introduces strategic planning, and this chapter will start from the research phase.

10.2 Research Phase

Before a space science mission is approved to enter the engineering development phase, it needs to go through several selections. What comes first is the selection into mission concept study, and the key technologies development of innovative payloads, which could be funded as two independent projects, led respectively by the science team and payloads development team.

Once a mission completes the concept study, it is qualified to enter the selection procedure for advanced research or enter directly into the intensive study, which is prior to the engineering development phase.

© Science Press 2021
J. Wu, *Introduction to Space Science*, Springer Aerospace Technology,
https://doi.org/10.1007/978-981-16-5751-1_10

Fig. 10.1 Flowchart of the full mission lifecycle of a space science mission

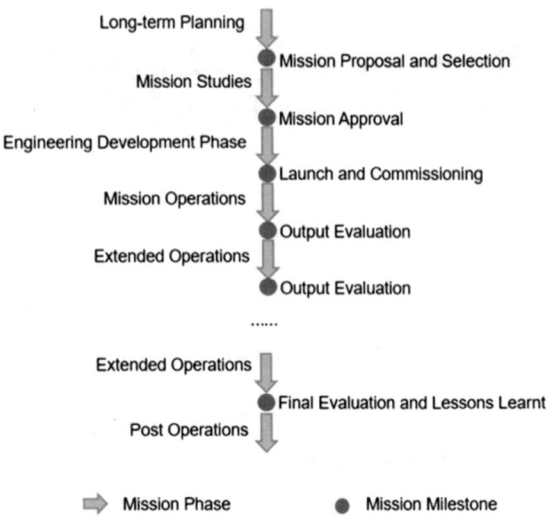

10.2.1 Mission Concept Study

The proposal of a space science mission always starts with a general concept. The mission concept study is funded in various science and technology programs in China, such as the National Key Research and Development Project from the Ministry of Science and Technology, the pre-study program from the State Administration of Science, Technology and Industry for National Defense, and the Strategic Priority Program (SPP) on Space Science from the Chinese Academy of Sciences (CAS). Generally speaking, these programs issue a call for mission proposals every 5 years. Among them, the SPP on space science issues a call for mission proposals every 2–3 years, with the budget for every mission 300,000–500,00 Chinese yuan, and research cycle 1–2 years.

For the concept study, it doesn't require a complete engineering design; instead, it focuses on the scientific objectives, payloads complement, mission description, and payloads' preliminary requirements for the spacecraft, among which the study of scientific objectives is the key point.

10.2.2 Advanced Research of Space Science Missions and Payloads

After the concept study is completed, the advanced research of space science missions and payloads will follow, covering the further study of scientific objectives and key payloads technology development. These projects also get funded through various programs. For example, the Advanced Research of Space Science Missions and

Payloads (hereinafter referred to as "advanced research") is funded within the framework of SPP on space science. And the call for mission proposals is issued every 2–3 years. Generally, the research period is 2–3 years, with a budget of 50,000–1,500,000 Chinese yuan per mission. It is worth mentioning that the budget may go up to several million yuan in the case of the great technology innovations of payloads involved.

As for the payloads with great technology innovations, it refers to those that can help yield new, innovative science data. If possible, the advanced research may fund its flight demonstration, such as the piggyback opportunity by sounding rockets and launchers. If there is no flight demonstration planned, it would be better for the payloads to undergo simulated space environmental tests in the ground labs. What needs to be pointed out is that the advanced research, aiming at the study of scientific objectives and the key payloads of technology research alike, should focus on scientific objectives, and should be led by a science team.

10.2.3 Intensive Study of Future Space Science Missions

The projects that complete advanced research have the possibility to be selected into the intensive study phase. Intensive study projects are also funded in the programs from the State Administration of Science, Technology and Industry for National Defense, and the Chinese Academy of Sciences. As for the SPP on Space Science of CAS, it issues a call for mission proposals for Intensive Study of Future Space Science Missions (hereinafter referred to as "intensive study") every 3–5 years, with the budget from several millions to dozens of millions yuan per project for a research period of 3–5 years. For some short, adaptable, and fast projects, the research period could be shortened to 1–2 years. In the intensive study phase, the further study of scientific objectives and payload key technology is the focus, and the mission management is research-oriented instead of engineering-oriented. Therefore, the projects are also led by the science team. However, compared with mission concept study and key technology research of payloads, the projects in intensive study have to meet more strict requirements in terms of mission schedule and engineering specifications.

Generally speaking, the projects in the intensive study phase should be the ones that have completed the mission concept study phase and key technology research phase of payloads, with the exception of some projects which are not included in the above-mentioned two phases but have already been technologically feasible after being funded by the individual home institute. The impact and involvement, as the two criteria of space science mission selection, are adhered to in the mission selection into the intensive study phase, resembling the selection process into the engineering development phase.

In the intensive study phase, scientific objectives are further studied the technology readiness level (TRL) of key technologies should be up to Level 5 or even higher and payloads' preliminary requirement for the spacecraft needs to be proposed. In the scenario that the project has the opportunity to fly its payloads, e.g., piggybacked by sounding rockets or satellites, its TRL could reach Level 6. Moreover, to identify the key technologies of spacecraft that are necessary for the implementation of the mission, some tests on the key technologies of spacecraft should be done if the available funding is sufficient. If the above-mentioned three aspects are carried out with high quality, and the impact and involvement of scientific objectives are great, then the project is likely to be officially approved to enter the engineering development phase. In the same line of reasoning, if the piggyback flight tests or ground simulation tests are not successful, the project needs to be terminated or applied for additional funding for further study and tests.

Therefore, it is not true that all of the projects in the intensive study enter into the engineering development phase directly. A final selection still needs to be done.

10.3   Reviews Necessary for the Approval

After the mission concept and key technology research phase, the intensive study phase, and the selection prior to the mission approval, a mission proposal is required to go through several reviews, such as review of scientific objectives and payload complement, review of payloads' preliminary requirements for the spacecraft including technology parameters, systems compatibility study including spacecraft, launcher, launching site, TT&C, ground operation, and scientific applications, and review of the mission budget. Figure 10.2 lists the work to be done prior to approval.

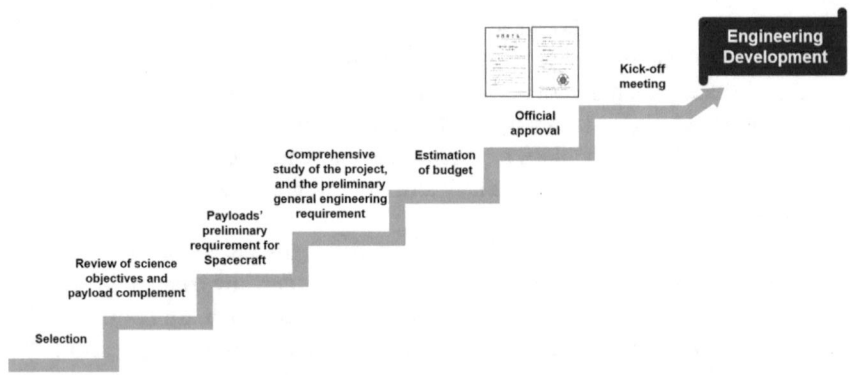

Fig. 10.2   The work to be done prior to the mission approval

10.3.1 Review of Scientific Objectives and Payload Complement

Although the impact and involvement of scientific objectives have been studied in earlier phases and the payloads have already been chosen, an official review is still necessary here, to freeze the scientific objectives and payloads complement. The drafting of the report should be led by the science team, with the participation of the payload team.

The point of the review is to see whether the scientific objectives are appropriate and whether they could be realized by the chosen payloads. To be specific, whether the objectives are too difficult to achieve or too specific to produce great impact, whether there is something missing from the scientific objectives, and whether the selected payloads are sufficient enough to address the scientific objectives.

This review is the first one in the engineering development phase, which sets the baseline of scientific objectives and payloads for the following engineering development. The report of the review acts as the preliminary reference for future engineering documents. Therefore, the scientific objectives and payloads are frozen once they passed the review, not allowing future changes unless there are insurmountable difficulties. What needs to be pointed out is that any changes would lead to risks in terms of technology, schedule, and budget.

10.3.2 Review of Payloads' Requirement for Spacecraft

The review of payloads' requirement for the spacecraft comes next. In the review, the size of spacecraft, mission's requirement for other systems, and the feasibility are scrutinized. The review should be led by the engineering team, with the participation of the science team. It is the most important phase for the engineering team and the science team to work together since any misunderstanding about mission definition and baseline could lead to duplication of design in the future and bring about risks. The review sets a baseline for the budget estimation in the next phase.

The review acts as input for the following engineering design. However, in view of the possible changes in the following key technology demonstration and design phase, the final solidification of the documents is about to be finished before the project enters the engineering qualification phase.

10.3.3 Systems Compatibility Study

With payloads' requirement for the spacecraft determined, the management agency organizes the review of six systems' compatibility, to compile the mission's engineering proposal and the preliminary general requirement for engineering development. The review, usually led by the chief designer to be appointed after mission approval, is an important step for overall engineering design, covering the determination of the launcher, launching site, TT&C, ground stations, etc. The report, together with the mission proposal, acts as the main technical documents supporting the mission approval [2].

10.3.4 Review of the Budget

The management agency needs to entrust a third party to estimate the budget, with an estimation report compiled. During the estimation, representatives of the related parties that propose the budget should be present at the meeting to answer questions posed by the estimation company. The estimation results, together with the report of six systems' compatibility, and the mission proposal, act as the important documents supporting the mission approval and the release of development permission.

10.4 Engineering Development Phase

The first thing after official mission approval is to appoint the persons in charge of mission engineering development. The persons in charge at the project level are the chief commander, chief designer, and Principal Investigator (PI).

The chief commander is fully responsible for the engineering development, including mission implementation schedule, quality, and budget. In view of the fact that participating entities cover institutes of the Chinese Academy of Sciences, universities, industry, launch site, TT&C system, etc., the chief commander needs to make sure that the project develops smoothly with the cooperation of participating entities. The science section and technology section is entrusted to the PI and the chief designer, respectively. If the chief designer cannot perform the duty, the chief command has the right to appoint a new person as the chief designer.

The participating institutes involved in the engineering development should respectively appoint the commander in charge of their own share of development, and it is usually a person with administrative duty since only the person with administrative duty has the necessary power of personnel appointment, resource allocation, quality control, and risk dissolution. To summarize, the chief commander and the commanders with administrative duty form the project's administrative structure.

The chief designer is responsible for the overall engineering design and tackling emerging technology challenges, and leads the designers at the system level and sub-system level.

What needs to be pointed out is that the PI is no longer the head of the project when the project enters the engineering development phase. As mentioned above, the head of the whole project in this phase is the chief commander. The PI is only responsible for the scientific objectives during the phase. But PI has a veto right once the engineering design and development deviate from the requirements of scientific objectives. If the PI decides to exercise the veto right, a report is needed to explain how the scientific objectives are jeopardized, which is followed by a review organized by the chief commander to decide whether to change the engineering design or terminate the project.

Up to now, the PIs of Chinese space science projects haven't exercised the veto right yet. But it is important to entitle PI with the veto right, to make sure that the engineering development of science missions should always serve the scientific objectives.

10.4.1 Preliminary Design Phase

After the mission approval, the engineering design team needs to go through the mission requirements again, including the re-study of key technologies and the feasibility, the necessary development of the experimental prototype, the verification of its physical properties, and the determination of technical baseline. The preliminary design phase is also called the experimental prototype phase.

If there are no insurmountable difficulties identified, the engineering design begins. The basic requirement of the design is to achieve a technical baseline through the most reliable, simplest, and most optimized method. Moreover, simulation tests at the spacecraft system level should be carried out when necessary. Besides, a review of the chosen components is needed to determine the type and model of all the components, and the grope test of components should be done when necessary.

The end of this phase is marked by the completion of engineering design and a review to solidify the general requirements for engineering development, which acts as the engineering input. That is to say, the science input is the above-mentioned scientific objectives and payload complement, while the engineering input is the general requirements for engineering development. The PI needs to attend all the reviews at the system level in this phase, and present clear viewpoints and suggestions. The engineering design completed in the phase is called "engineering qualification design", which needs to pass the formal engineering qualification design review taking place in the late preliminary design phase or the early engineering qualification phase. Next, the project enters the engineering qualification phase.

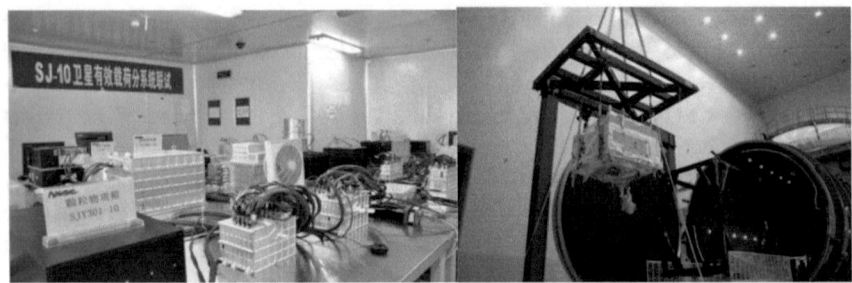

Fig. 10.3 SJ-10 mission's joint desktop test of payloads and DAMPE mission's thermal vacuum test at a spacecraft system level

10.4.2 Engineering Qualification Phase

In the engineering qualification phase (or qualification model phase for short), the engineering team completes engineering development based on the design and conducts environmental tests with condition parameters 1.5 times harsher than the real space environment, which includes but is not limited to mechanics vibration test, thermal vacuum and thermal cycling test, and the electromagnetic compatibility test. Figure 10.3 shows the SJ-10 mission's joint desktop test of payloads and the DAMPE mission's thermal vacuum test at a spacecraft system level.

Although the engineering prototype developed in the phase will not be launched, the development must meet the launching standard, because the purpose of this phase is to discover and solve emerging problems. If there are no problems emerging, the engineering design will not be changed. However, once there is a problem arising, the engineering design needs to be adjusted, which needs to follow the necessary procedures.

This phase ends after the engineering prototype passes all the environmental tests, the development report is completed, and the final flight model design report is compiled. The PI of the mission needs to attend all the reviews at the system level in this phase, and present clear viewpoints and suggestions.

10.4.3 Flight Model Production Phase

The flight model production phase aims to develop the flight model to be launched. To verify whether the manufacturing is correct, the flight model, as the product, needs to undergo environmental tests one more time, which are the tests at acceptance level. However, the magnitude of the tests is reduced to half compared with the previous environmental tests, to guarantee the safety of the product. In this phase, the control of the technical status is stricter, and no more changes are allowed unless a new

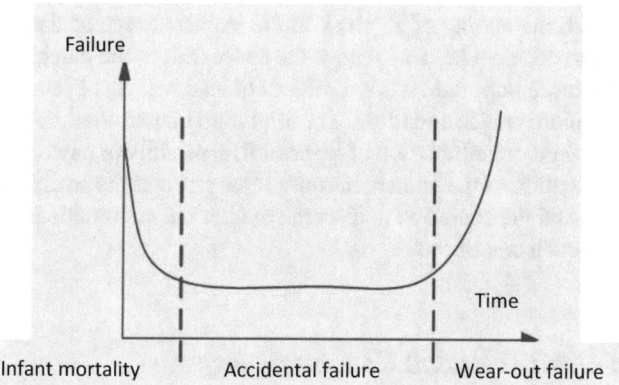

Fig. 10.4 Bathtub curve of components

technology problem emerges. The technical status needs to be exactly the same as the status determined in the engineering qualification phase.

In the late phase, a reliability growth test needs to be conducted, due to the bathtub curve that indicates the failure period of components (Fig. 10.4). To make components pass their infant mortality period, an additional in-circuit test of the electrical equipment is needed. As to the time length of in-circuit test, it needs to be determined according to the difference between the time length of the test after component mounting and the time length of the infant mortality period. Usually, several hundreds of hours are needed for the in-circuit test after all the spacecraft system-level tests are done.

The completion of the development report, together with passing the factory acceptance reviews, marks the end of flight model production phase. The factory acceptance reviews include the review at the spacecraft system level and the one at the project level. The review at the project level is organized by the management agency. The spacecraft system, together with the other five systems, needs to attend the review meeting to state its readiness for launch. The PI also needs to be present at the meeting, to give clear decisions on the factory acceptance review.

10.4.4 Tests and Launch

After the factory acceptance, the spacecraft is transported to the launch site; hence entering the last phase prior to the launch. During this phase, no more environmental tests are to be carried out. The only tests required are the various electricity tests similar to those before factory acceptance. However, due to the concern for safety, fueling is usually done at the launch site for those satellites that need fuel. The fueling is actually a high-risk job in this phase. After the fueling, the leakage and mass center test follows. With the ending of all work in the technical area of the launch site, a review follows to decide whether to move the spacecraft to the launch pad.

There is one more important work for the PI to lead before the launch, i.e., determining the minimum launch condition. The minimum launch condition aims to identify the possible fault or malfunction of spacecraft, especially of payloads, which will lead to the termination of the launch, because it jeopardizes the realization of scientific objectives. And the condition to discern whether the malfunction will happen is the minimum launch condition.

10.4.5   In-Orbit Tests and Commissioning

After the launch, in-orbit test of the platform in spacecraft system is to be conducted at first, such as the power system, the attitude control system, and the tracking telemetry and control system, which is followed by the in-orbit test of payloads, and the calibration and setting of various parameters. The in-orbit test of payloads is organized by the science team under the leadership of PI.

After the above-mentioned tests are finished, the spacecraft is formally commissioned to the TT&C system, mission operation system, and science application system.

References

1. Wu J, Fan QL, Cao S et al (2018) Progress of strategic priority program on space science. Chin J Space Sci 38(5):1–6
2. Wu J, Giménez A (2020) On the maximization of the science output of space missions. Space Sci Rev. https://doi.org/10.1007/S11214-019-0628-4

Chapter 11
Management (III): Quality Management and Risk Control

11.1 Introduction

A space science project is a space engineering project in the first place. The realization of its scientific objectives depends on the successful mission implementation. Hence, the quality management and risk control of a space science mission basically follow the same principles as with other space missions. For a science team that participates in a space project for the first time, the team members need to be aware of the fundamental points of quality management and risk control and learn how to communicate with the engineering team. This chapter introduces the principles of quality management and risk control in a space science project, with the members of the science team as the potential readers, and assuming that the spacecraft team is already very familiar with the engineering development.

11.2 Quality Management

Quality management is a general term for a variety of management work during production to ensure the quality of products. The systematic construction of China's quality standard system starts in the 1990s, and the current generally accepted quality standard system is quality management system requirements (GJB 9001C-2017). All the entities involved in the product development of a space science mission need to pass the certification of the entrusted certificate authority, to adopt the same quality management standard.

The quality system has three core characteristics, i.e., being user-centered, all-staff participation, and constant improvement.

To start with, the production needs to be user-centered. For a space science mission, the "center" is scientific objectives proposed by the science team led by the PI. However, when it comes to a specific product, the user is who actually operates it. For example, the user of a satellite transponder is the spacecraft team, and the

© Science Press 2021
J. Wu, *Introduction to Space Science*, Springer Aerospace Technology,
https://doi.org/10.1007/978-981-16-5751-1_11

immediate user of payloads is also the spacecraft team. The reason behind this is that the spacecraft team will represent the science team to take care of the science user's requirements of the payloads' technical parameters which derive from the scientific objectives. It is also because that the payloads can perform their functions to acquire data only after they are successfully installed on the spacecraft. While meeting the technical requirements formulated by the spacecraft team, the entity responsible for the payload development should keep in close touch with the science team to make them aware of the whole development process, in view of the fact that they are the final data user. Once the science user poses a question of concern, the institutes that develop payloads have the obligation, on behalf of the science team, to suggest adjustments to the spacecraft team. However, the spacecraft team has the final say on the adjustment. Any subsystem cannot make an independent decision because any changes or adjustments at the subsystem levels could lead to systematic re-structure, which would be introduced in detail in the following text about technical status.

In addition, all-staff participation means that all the departments involved in the development process in an entity should participate in the construction of the quality system, not only including the departments that actually develop the products but also the administrative departments and the departments providing the logistic support. This is because the quality of products is related to all the staff of the entity. For example, the training and appointment of the development team need the authorization of the human resource department; the administrative office is responsible for the effectiveness and consistency of appointment files and other management files; the assets department is in charge of the procurement of raw materials and components; and the logistics department is responsible for products' transportation. Therefore, the responsibility of quality control definitely rests not only on the shoulders of the development team.

Finally, the construction of a quality system is not a one-step-to-reach business but needs constant improvement. All the entities that have established a quality management system started from scratch, and have been optimizing it constantly. It is not an one-step-to-reach business, not only because the entity is evolving and the staff is improving by continuous learning, but also because the external environment is changing, including the overall quality of personnel (e.g., new employee recruited), the changes of raw material and components quality, the changes of user's requirements, and changes of weather conditions during product transportation. What needs particular attention is that according to the system's requirement, the product quality and related service needs the feedback from users, which drives the quality system optimization by the product development entity.

It is true that the prototypes developed during the advanced research phase and intensive study phase of a space science mission only provide functional verification, and are not formal products, which means they may not be managed according to quality management system requirements. However, to ensure the high quality and traceability of science research, a tailored quality management method could greatly improve the development schedule and reproducibility of functional prototypes. Once a space science mission enters the engineering development phase, especially the engineering qualification phase and flight model production phase, it is a basic requirement for the entities involved in the development to establish a quality system according to GJB9001 and pass the above-mentioned certification.

The fundamental aspects of quality management are introduced as follows.

11.2.1 Quality Manual and Procedure Documentation

Quality manual is the top-level quality management document for the entities involved in the development, finalizing its quality management principle, goal, structure and method, and the responsibilities of its administrative leaders and related departments. The quality manual is signed for release by the legal person of an entity (head of the entity).

The procedure documentation belongs to the second level, standardizing the main management procedures. The second-level documentation is also relatively mature and is usually open to the public.

In order to streamline the development work, a third-level documentation is usually needed to formalize the detailed working methods, which is the innovative documentation adaptive to specific conditions of an entity. When a quality supervisor performs verification in an entity, a confidential letter should be signed to guarantee that the third-level documentation will not be disclosed.

11.2.2 Documentation Control

Document control is one of the core businesses of a quality management system. The management documentation and technical documentation alike should be subject to strict control, covering the approval files before the release, related changes and adjustments, designation and distribution, termination and invalidation, etc. to ensure the validity of documentation.

In terms of technical documentation control, in a complex engineering project, in view of the fact that no one can do all the technical work, it is necessary to break down the work, which means dividing the system into subsystems, the subsystems into smaller subsystems, etc. The connection between the hierarchy of systems and subsystems is called the "technical interface connection", which needs to be formalized through documentation. During the development phase, changes of technical status often happen, making it necessary to update the corresponding interface documentation. Since the documents are all distributed to relative departments and subsystems or smaller subsystems, and once the updated documents are not strictly controlled, failing to ensure its validity, the problem of inconsistency in the technical interface would arise, leading to quality problems.

In terms of management documentation control, for example, if the validity of an appointment file of a designer failed, such as the appointed person in the file is no longer in charge, it will lead to quality problems caused by unclear responsibilities.

Therefore, to determine whether an entity has successfully established a quality system and whether the system operates well, a basic indicator is to see whether the entity achieved validity of documentation control.

11.2.3 Closed-Loop Solution of Quality Problems

Various quality problems are inevitable in space product development and manufacture, especially in the engineering qualification phase, during which problems usually arise in the tests. However, the arising of problems is not necessarily a bad thing, but actually the purpose of the tests. Once problems arise in the tests, we should analyze from technical and management perspectives, and then adopt necessary measures to avoid problems showing up again. The process to thoroughly solve the problems is called "closed-loop solution".

For the Closed-Loop Solution of Technical Problems

The five requirements for closed-loop solution of technical problems are the following: *Where, Why, Repetition, Solution*, and *Lessons learned.*

Where means to identify the accurate root cause of the problems. For example, we need to trace down to the specific components if something is wrong with the circuit board.

Why means to determine the physical and chemical mechanism of the quality problems through tools such as theoretical analysis or tests.

Repetition means to repeat the process that caused the quality problems through methods such as tests and other verification tools to verify the above-mentioned root causes and mechanism determination.

Solution means to adopt effective measures to correct the quality problems and further ensure the effectiveness of the measures taken through tests in a bid to thoroughly solve the problems.

Lessons learned refers to sharing the quality problems with all the entities involved in mission model development, even including those entities involved in other mission models, to examine whether they have similar problems with a similar mechanism so as to take necessary precautions.

For the Closed-Loop Solution of Management Problems

Sometimes quality problems arise in the development phase due to management misconduct, which is called "management problems". There are five requirements for the closed-loop solution of management problems, namely *How, Who, Solution, Responsibility*, and *Complete regulations.*

How means to examine the whole process where problems arise and develop, to find out the weak link or management loopholes.

Who means to identify the accountable entities and persons, and distinguish the major and minor responsibilities.

Solution means to work out and adopt corrective and preventive measures against the weak link and management loopholes.

Responsibility means to treat seriously the quality problems due to management errors so as to improve management. In the case of repetitive quality problems and problems due to misconduct, it is necessary to seriously punish involved entities and persons according to relevant regulations. If the problem is very severe in nature, the accountable personal should be removed from the position or even dismissed.

Regulation Improvement mean to improve and exercise regulations against the weak link and management loopholes. In systems engineering, especially in space systems engineering, almost all the technical quality problems are accompanied by some management problems, even serious management loopholes. Therefore, in actual practice, usually both closed-loop solutions of technical and management problems are required.

11.2.4 Technical Status Control

Depending on the product development phases, the technical status is divided into functional baseline (design phase), development baseline (engineering qualification phase), and product baseline (flight model production phase).

Functional baseline usually finds itself in the documents such as approved mission proposals and general technical specifications and requirements.

Development baseline finds itself in the documents such as approved design documents and test conditions.

Product baseline usually finds itself in technical documents such as approved design documents, design drawings, process documentations, and test standards.

The three baseline documents proceed one after another in a coordinated manner and are characterized by traceability. Once the baseline is settled, the control of its change is one of the core businesses of quality management. The change of technical status falls into three categories.

Category III represents the highest rank, which includes the changes of system-level technical status, changes of interface connection between systems, and changes of interface connection between subsystems, or the changes of subsystem's performance, parameter, reliability, and external form, or the changes of design parameters of key components.

Category II includes important changes, covering the changes of interface connection in subsystems, or changes of instrument-level technical parameters, which do not affect the parameters of the subsystem.

Category I includes ordinary changes which do not affect the performance parameter and the interface of other relevant parities.

Any changes involving technical status need to pass the review, which is characterized by five requirements, i.e., *Study, Consensus, Verification, Approval,* and *Implementation.*

Study means to carry out a thorough study of the necessity for the change, the technology replacement, and the implementation scheme.

Consensus means that the implementation scheme needs to be confirmed and signed by all the involved systems, subsystems, and thesmaller subsystems.

Verification means to verify the implementation scheme through tests, to prove the changes are effective and implementable.

Approval means that changes need to be approved by a higher-rank person in charge of the subsystems\systems, depending on the category of the change.

Implementation means that all the changes which pass the review and get approved should be executed according to the technical implementation scheme, and relevant technical documents should be updated accordingly so as to form the closed-loop management.

For a space science mission, it is of great importance to keep the technical baseline unchanged. Once the change is inevitable, the above-mentioned five requirements need to be met. The PI and his\her team often pose new demands when they want to improve the mission's function and performance after the project enters the engineering development phase. In this scenario, if the procedure of changing the technical status is not strictly followed, a major hidden quality risk may be sown. This is why the payload development team needs to work under the leadership of the spacecraft team, and why changes should be made by strictly following the five requirements for space engineering. The payload development team is forbidden to take the liberty to make changes after the science team poses the new requirement.

If the science team constantly poses requirements of change in the engineering development phase, then it means the mission team has not done a thorough study during the advanced research phase, intensive study phase, and the comprehensive study phase before official approval. In principle, the requirements from the science team should be determined during the review of scientific objectives and payload complement, and the review of payloads' requirements for spacecraft development prior to the mission approval. Of course, it doesn't mean that there is no chance for changes after entering the engineering development, but all the changes need to follow the management procedures for quality control mentioned above. The science team needs to be aware that any change in the engineering development phase will lead to a dramatic increase in cost and risk.

11.3 Risk Control

A space science mission, like other space missions, is a high-risk systems engineering project. The risk is not only from the launching process but also due to the fact that the spacecraft and its products have almost no chance for repair once they are launched into space. Moreover, in-orbit spacecraft also has the possibility to experience malfunctions due to space weather events caused by the solar eruption.

Risk is defined by the odds of happening and the influence after happening. In terms of the odds, a space science mission usually has greater odds of risk. It is because that a space science mission adopts more new technologies to achieve scientific breakthroughs. In particular, the more advanced the payloads, the lower the technical readiness level (TRL). In terms of the influence, a space science mission has a relatively smaller influence compared with other types of space missions, in that science research is an exploration in nature and the public are more tolerant with its failure or partial failure. However, the government is less tolerant because the budget of a space science mission has a large share in basic science. Thus success is more desirable on the government side, in view of the huge amount of money invested. Taking the above-mentioned two factors into consideration, strict risk control and management is needed for a space science mission.

The following is a brief introduction of several risks and corresponding control methods.

11.3.1 Risks Identification and Prediction

Technical Risk

Technical risk mainly comes from key technologies of the spacecraft and the payloads, while the potential risk from the new technologies of payloads is higher. These technologies not only include hardware but also data analysis methods, software, etc., which will directly affect the science output.

Product Quality Risk

Product quality risk usually comes from the entities involved in the engineering development of a space science mission, especially those which are involved, for the first time in the payload development. They are less experienced in systems engineering and tend to treat the management of engineering development the same as the basic research projects. Their documentation control, technical control, and management of closed-loop solutions may not be fully executed, which may lead to product quality risk.

Risk of Responsibility and Schedule

Breaking down the responsibility to individual persons is one of the core methods of systems engineering management. However, the personnel flow in basic science research is big. A space science mission takes 10 years or even longer from the research phase to the launching and operation. Therefore, compared with other institutions, there is a bigger personnel flow in the entities involved in payload development, and the entity to which the science application system is affiliated, which may lead to risks of blurred responsibilities, schedule delay, etc.

Risk of International Cooperation

International cooperation is encouraged in space science, which helps to broaden the view, keep abreast of advanced technologies and management practices, expand the science team, and increase science output. However, we cannot ask the international partners to use the same management procedures as our domestic entities, since different countries have different management procedures. In the event of schedule delay, technical status change, and closed-loop solutions to technical problems, we are left in an awkward situation.

Risk of the Components Import

In view of the fact that payloads of a space science mission have high technical parameters, it is necessary, in most cases, to import advanced components. However, China faces the challenge of technology blockade, such as the *International Traffic in Arms Regulations* (ITAR) issued by the USA in the 1990s. To make the situation worse, a small number of components available to China are very expensive and there are harsh commercial terms attached. All of these may lead to schedule delay and quality risk.

11.3.2 Risk Control and Management

For the Technical Risk

In the advanced research phase, the correctness of principles related to technology should be assured. The intensive study phase mainly aims to develop engineering key technologies, improve the design, and integrate key technologies into a whole design for verification. In the preliminary design phase, the focus is on the material-ization of the physical functions, which will be verified through a complete electronic performance prototype. In the engineering qualification phase, the correctness of the design should be fully tested in a simulated space environment to expose the potential problems, which will be solved accordingly. In flight model production phase, the focus is on the technical process where the risk of safety issues and malfunctions due to the application of new techniques will be eliminated. During the whole development process, the problems about principles, design, test, and production should be reviewed constantly to connect the risk control and the corresponding solutions of quality management problems in every phase so as to make a complete product history.

For the Product Quality Risk

First of all, the entities involved in the development should be supervised to establish a quality system and pass the certification. It takes 1–2 years for even a department in an entity to establish the quality system independently from scratch. Therefore, it is necessary to arrange training as early as possible and the construction of a quality system should be supervised. The deadline to pass the certification could be set on a date prior to the engineering qualification phase, during which the development of products needs to be fully under control, and there are strict requirements for the changes of documentations and technical status, as well as for the closed-loop quality problem-solving. The entities need to comply with higher system-level management, using the same quality management language and requirements.

For the Risk of Responsibility and Schedule

According to the requirement of human resource management, the responsibility should be broken down to individual persons. When there is a personnel change, it is necessary to appoint and train new staff to avoid the vacancy of responsibility. Meanwhile, the training of special professionals should be strengthened to avoid the situation that newcomers cannot meet the requirements.

For the Risk of International Cooperation

First of all, the partner with a better degree of mutual trust should be chosen, such as those with whom there is sound shared cooperation history or those who are guaranteed by intergovernmental agreements. China and ESA have been maintaining a friendly partnership with mutual trust. Even during the Severe Acute Respiratory Syndrome (SARS) epidemic outbreak in 2003, the schedule of Double Star Program payloads development by ESA and the following acceptance review were not affected. Secondly, the partner's working style and schedule keeping should be taken into account when drafting the mission development plan to leave sufficient margins, and the mission schedule should be coordinated through joint review. When an unexpected scenario arises on the partner's side, it is necessary to dispatch staff to the partner's institution for onsite coordination so as to keep the development schedule.

For the Risk of the Components Import

Compared with the procurement of domestic components, the components import channel should be settled earlier and the order should be placed as soon as possible. In the design phase, the backup plan (Plan B) for the procurement of domestic components or imported components should also be considered. Both plans should be kept before the components are safely in hands so as to avoid the engineering schedule delay.

Chapter 12
Management (IV): Full Mission Lifecycle Management and Output Evaluation

12.1 Introduction

The fundamental difference of a space science mission from other space missions lies in its output. For space missions supporting market-driven applications, the assessment of performance is clearly defined by the quality of the service provided and its added value, such as communications, TV broadcasting, navigation, weather forecast, ocean forecast, etc. For space science missions, success is evaluated in terms of the advancement of scientific knowledge, based on new discoveries and the tests of the law of nature. From the perspective of management, a space science mission not only needs to ensure its success of mission development and launching but also to guarantee the science output. To optimize the output, the management agency needs to take the full mission lifecycle into consideration, from the mission study, to selection, and to the operations after launch. In this chapter, some management points over the whole lifecycle of a space science mission will be discussed.

12.2 Relationship of the Stakeholders of Space Science Missions

In order to clearly identify the key points in the full mission lifecycle management of a space science mission, it is necessary to analyze first the roles of all the parties involved [1] and their relationships.

The most important stakeholder is the government. The government investment is the main source of mission budget, in spite that there is a small number of private enterprises willing to participate in some science missions.

The government investment in space science missions mainly comes from fiscal appropriation, i.e., the government revenue from taxpayers. Generally speaking, the investment needs to be materialized through the budget, after the government's budget proposal gets approved by the national legislative department, e.g., NASA's annual

© Science Press 2021
J. Wu, *Introduction to Space Science*, Springer Aerospace Technology,
https://doi.org/10.1007/978-981-16-5751-1_12

Fig. 12.1 China's
government investment in
R&D, basic research, and
space science (the size of the
circles do not reflect the
actual shares)

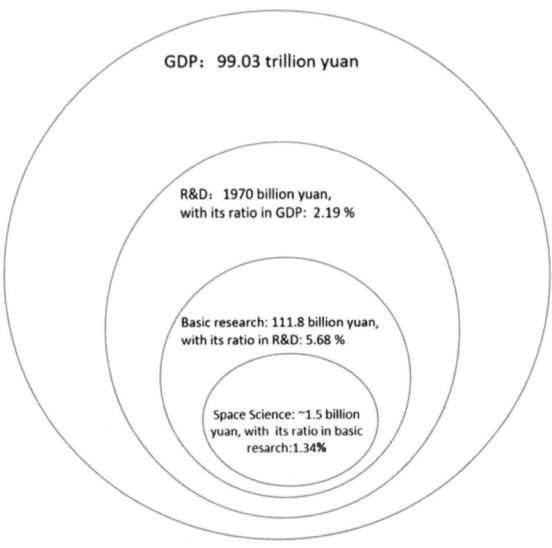

GDP: 99.03 trillion yuan

R&D: 1970 billion yuan,
with its ratio in GDP: 2.19 %

Basic research: 111.8 billion yuan,
with its ratio in R&D: 5.68 %

Space Science: ~1.5 billion
yuan, with its ratio in basic
research:1.34%

budget needs to be approved by the U.S. House Commerce, Justice, Science Subcommittee, while in China, the government budget needs to be approved by National People's Congress. In China, the ratio of R&D (research & development) spending in GDP and its growth rate need to be discussed and approved by the National People's Congress. Taking the year 2018 as an example, the ratio of R&D in GDP in China is approaching 2.2%, approximately 1,970 billion yuan. With the growth of the national economy, the ratio is going to increase up to around 2.5% and then get stabilized. The ratio in countries like Switzerland and South Korea has exceeded 3%. In 2018, China invested 5.68% of its R&D spending in basic research, approximately 111.8 billion yuan. Out of basic research, around 1.34% was invested in space science satellite missions, approximately 1.5 billion yuan, as shown in Fig. 12.1.

Although taxpayers, or the public, do not play the role to directly approve the projects, the public pays great attention to the output of government investment in basic research; meanwhile, the public expresses opinions on it. For example, the DAMPE (Wukong) mission and QUESS (Mozi) mission launched during the 12th Five-year Plan period have attracted high public attention.

The science team that proposes a space science mission and analyzes science data is a special group belonging to the public and plays an essential role in the mission. They are part of the public, and also part of the execution team of a space science mission that is responsible for the science output.

Aside from the government department and management agency, other stakeholders are the entities involved in mission development and the research institutes responsible for the scientific research. The entities involved in mission development include the state-owned aerospace industry, as represented by China Academy of Space Technology, Shanghai Academy of Spaceflight Technology, and China Spacesat Co., Ltd., and some private aerospace enterprises. The entities

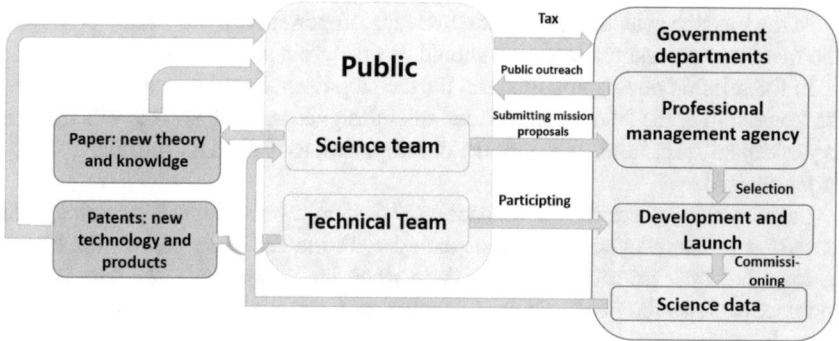

Fig. 12.2 Relationship of the stakeholders

involved in scientific research mainly include research institutes and universities, i.e., those proposing the mission and participating in data analysis. Figure 12.2 is the relationship chart of the stakeholders.

A space science mission proposal should come from the science community, i.e., the mission should be proposed following the *bottom-up principle*. The reason behind is to maximize the science output. The mission team would be more motivated in data analysis if the mission is proposed by themselves, while their enthusiasm would be compromised if the mission is proposed by the government. It is especially the case for the PI, who plays a crucial role in the team. The PI is less enthusiastic about the mission if the mission is approved before his\her appointment. Therefore, the *bottom-up principle* is the first step to maximize the science output. Space science missions are proposed in this way across the globe.

In China, the government departments in charge of space science projects mainly include the Ministry of Science and Technology (MOST), the Chinese Academy of Sciences (CAS), the State Administration of Science, Technology, and Industry for National Defense, etc. These departments entrust management work, such as planning, selection, development, launch, and operation, to a professional public institute after the budget is approved. China has three professional public institutes/management agencies in charge of space science management, namely the China Manned Space Engineering Office, Lunar Exploration and Space Engineering Center of CNSA, and the National Space Science Center of CAS.

The professional management agency plays a very important role in full mission lifecycle management. It organizes scientists for strategic planning, issues a call for missions, and organizes mission selection, mission development, launch and operations, and the output evaluation. During the whole process, the basic principle must be preserved, i.e., to maximize the science output. The specific measures during each phase are as follows:

In the strategic planning phase, the planning should aim at science frontiers.

In the mission proposal phase, the missions should be proposed following the *bottom-up principle*.

In the mission selection phase, the two most important criteria should be impact and involvement, and the selection should be open, fair, and unbiased.

In the mission development phase, the development should be carried out under the leadership of the "chief commander/chief designer and PI", and the PI's rights is to participate, track, and supervise the development, as well as its veto right, should be guaranteed.

In the mission operation phase, necessary data policy should be formulated, such as setting up the exclusive period of data use. In the exclusive period, the PI and its team have the priority to use the data; when the period ends, the data is open and accessible to the public so as to involve as many scientists as possible in data analysis.

In the phase of mission extension, evaluation, and engineering summary, it is necessary to carry out an overall evaluation of science output and report to the government and public with clear and quantitative results.

The details of the first five phases have been introduced in previous chapters. The following is about the specific requirements in the evaluation phase.

12.3   Output Evaluation

Output evaluation is the last link of full mission lifecycle management, and also the most crucial one since it is the feedback to government investment. The feedback includes nation's honor, science output, new technology and products, and public outreach, etc., which helps to strengthen national competitiveness, improve human knowledge about nature, enhance spin-in and spin-out of space exploration, and inspire the young generation to engage in science, technology, engineering, mathematics (STEM) study. We can declare the full mission lifecycle completed only after the public gets the feedback, as shown in Fig. 12.2. Space Science endeavor is sustainable only when the feedback is positive, which is to say, the output is recognized as worthwhile by the public.

When is the appropriate time to evaluate output? Generally, the first evaluation is carried out at the end of the mission's designated lifetime. Scientists may think it is too early since data analysis takes time. Starting from the commissioning of a space science mission, the number of papers using its data goes up slowly, then gets stabilized somewhere, and then declines after several years. During the designated lifetime, it does not indicate the peak of science output, especially for those missions with a shorter designated lifetime. The management agency in charge of the evaluation should be aware of this.

There are two scenarios when it comes to the end of the designated lifetime of a space science mission. The first is that the mission has to be terminated due to its failure, with scientific objectives unrealized. The other scenario is that the mission operates well, which is the most common case. Nearly all scientific satellites worldwide enter mission extension after the end of their designated lifetime. An evaluation of output is needed to see whether it is necessary to enter mission extension. If a mission has accomplished its scientific objectives, or even better than the expectation, the user will apply for extending the operation. An extension means more budget. Hence two applications are necessary, namely application for extending operations and application for adding budget. As to the evaluation, the already achieved science output and other aspects should be evaluated.

Evaluation of output covers three aspects. The first one, also the most important one, is the science output [2]. Science output is evaluated mainly by the papers published in academic journals with independent peer-review procedures. The evaluation could be carried out against two metrics, namely the impact factor of the journal in which papers are published, and the quantities of papers. The former implies the impact, while the latter implies the involvement, which are the criteria of mission selection. A prolific space science mission yields over 100 papers per year. Taking the SWIFT mission as an example, around 300 papers are published per year with its data. Another example is HST, over 10,000 papers have been published using its data since its launch in 1990. The joint data analysis of DSP of China and cluster of ESA yielded around 100 papers per year.

The second aspect is the results of technology transfer. The motivation for a nation's investment in a space science mission includes making breakthroughs in the science frontier, as well as the development of high technology. A space science mission adopts a great number of new technologies. However, it is not the main goal of the science team to transfer the technology, and it is the management agency that should take it into account. Once entering the engineering development phase, the agency should analyze possible new technologies that the mission would bring along, pay attention to the patents that would be produced as the mission development progresses, and encourage the engineering development team to apply for patents. When the mission ends, the agency should make an evaluation of technology transfer. Both NASA and ESA set up specific offices in charge of the technology output and technology transfer, which follow the mission since the official mission approval, apply for patent registrations if new technologies are identified, make an assessment whether these technologies could be transferred, and even allocate special funds to incubate businesses. In addition, if there are market-oriented applications, it would be better to count the product value. For example, European Organization for Nuclear Research (CERN) signed procurement contracts with about 100 enterprises with nominal values of around 20 million euros in 2007. 70% of the 100 enterprises have developed new products, and 50% of the enterprises have value-added products due to new products. What needs to be pointed out is that while evaluating the technology transfer, science output should not be neglected. If a mission only produces technology output, it cannot be claimed as a science mission. It is wrong to only focus on technology development in the name of space science, misleading and abusing the national investment in the basic science sector.

The third aspect is the influence of a space science mission on the public. As mentioned above, government investment in space science mainly comes from national fiscal appropriation, i.e., from taxpayers. Therefore, the output should be delivered to the public. The USA has done an investigation, which shows that the input-output ratio of the Apollo program is 1:7; the EU has also done a similar investigation, which shows that 15 euros are yielded with 1 euro invested in the space sector. It needs to be pointed out that the input-output ratio of 1:15 counted by the EU covers not only the science sector but also navigation, communication, meteorology, etc. Besides, the public outreach and aspiration triggered by space missions are more difficult to evaluate. The agency needs to let the public be familiar with the scientific objectives of a space science mission upon its official approval, even to brief them about the development schedule, and to introduce popular science to them. The fundamental knowledge about the universe and space is the hot spot of the public's interest, especially for the young generation. Breakthroughs in the area are very helpful to inspire the public and the young generation to engage in STEM study [3]. We can see that the openness of a space science mission is a great advantage to promote public outreach, which is a public duty for the management agency and entities and teams involved in mission development. To evaluate the output in this aspect, during the mission development or when the mission ends, is also an important work in full mission lifecycle management.

If the evaluation result shows that the mission is possible to yield more output, the application for extension should be approved so as to maximize the output. When the extension ends, the agency needs to evaluate again the output during the extension, all the way through to the end of the mission. When a mission ends, a comprehensive evaluation and summary of engineering development should be carried out [4].

References

1. Cameron BG, Seher T, Crawley EF (2011) Goals for space exploration based on stakeholder value network considerations. Acta Astronaut 68(11/12):2088–2097
2. Wu J, Bonnet R (2017) Maximize the impacts of space science. Nature 551:435–436
3. Gonzalez HB, Kuenzi JJ (2012) Science, technology, engineering, and mathematics (STEM) education: a primer. Congressional Research Service, Washington
4. Wu J, Giménez A (2020) On the maximization of the science output of space missions. Space Sci Rev. https://doi.org/10.1007/S11214-019-0628-4

Chapter 13
International Cooperation

13.1 Introduction

Since the beginning of the space age, space science and technology is primarily used as the tool for national politics, which has something to do with the fact that space science and technology is closely related to national defense. There is no denying that space science and technology is actually derived from national defense technologies, as shown in the case of the V2 rocket. Or put it alternatively, since the birth of space science and technology, it was first applied in the national defense.

The curiosity of human beings to explore the unknown is boundless. Since Galileo, out of curiosity, pointed his telescope to space, the science and technology in human history has made a giant leap forward. The Soviet Union launched the first lunar probe in 1959. In the 1960s, the Earth observation satellite missions and the scientific missions to explore the cosmos began to flourish.

When a space mission evolves into a scientific mission, international cooperation will become ievitable. The reason behind this is that science knows no boundaries, and the mansion of human scientific knowledge is jointly built by people bit by bit across the globe. In the 1960s, some European countries first joined hands together and established the European Space Research Organization (ESRO), dedicated mainly to scientific explorations, which now evolves into European Space Agency (ESA) [1]. Following that, the world has witnessed increasing international cooperation, such as Europe–USA, Europe–Soviet Union/Russia, Europe–Japan, Europe–China, USA–Soviet Union/Russia cooperation. It is fair to say that international cooperation in the field of space happens first on space science program and space science is currently the mainstream of international cooperation.

This chapter will touch upon the necessity for international cooperation, its main cooperation modes, and the challenges.

What needs particular attention is that the international cooperation mentioned in this chapter is intergovernmental cooperation based on no-fund exchanges. The cooperation that involves procurements is deemed as commercial and will not be dealt with in this chapter.

© Science Press 2021
J. Wu, *Introduction to Space Science*, Springer Aerospace Technology,
https://doi.org/10.1007/978-981-16-5751-1_13

Fig. 13.1   The major space agencies responsible for the implementation of space science missions

The major space agencies responsible for the implementation of space science missions are listed in Fig. 13.1.

13.2   Necessity for International Cooperation

It is self-evident that the objective of space science missions is to study and understand nature and the universe. It strives to build models and test their validity by carrying out observations. The planet Earth where we are living, other planets in the solar system, the stars, galaxies, and the cosmos at large are all the targets of our curiosity-driven explorations. The objective does not belong to the scientists of an individual country. It is the common objective for all scientists all over the world. Therefore, international cooperation in the field of space science is, by no means, only a tool. It is a joint science activity for us to better understand nature.

In order to coordinate and promote international cooperation in the field of space science, the International Council of Scientific Unions (ICSU) established the Committee on Space Research (COSPAR) in the year 1958 in the early days of space age. COSPAR is the only international organization that covers all the disciplines in space science and has thousands of associates across the globe and dozens of institution members. It hosts a biennial COSPAR Scientific Assembly on even years to promote academic exchanges and international cooperation among different countries in the field of space research.

The governments' motivation for international cooperation is multi-fold. First of all, for the scientific frontier discovery and breakthrough, only the No. 1 is heralded and fully acknowledged. When the individual government is striving to be No. 1 as their political objective, they also need to avoid the duplication of what has been done or what is already planned by other countries. Therefore, all the government departments in charge of space science missions require the scientists, from the strategic planning phase, to make mission proposals by considering the international scientific frontiers. Inviting important international cooperation partners to attend the strategic planning workshop is the common practice, in a bid to coordinate their own space science missions to avoid duplications. When several countries are interested in one specific research, it is possible to reach a consensus by the parties involved in terms of labor distribution to make a coordinated effort for a breakthrough. A perfect example is the Inter-Agency Consultative Group for Space Science (IACG) established by NASA, ESA, and space agencies in the Soviet Union and Japan when comet 1P/Halley returns to the inner solar system in 1986. The involved space agencies launched six spacecraft to 1P/Halley for joint detection from different angles, different distances, and with different instruments.

Secondly, the individual country's budget for space science is limited. For a specific country, a large number of mission proposals from the scientists have to go through repeated selection procedures. In this process, some good ideas may be dropped, and even the flight opportunities for many innovative instruments are missed. Therefore, in the cooperation practice, the national space agencies unanimously take advantage of the launch support and rideshare opportunities as offered by collaborative countries. For example, in the scenario of bilateral cooperation, one country takes the leading role to develop a satellite, while the other country provides the launch opportunity, and then the data is shared. Another scenario is that the mission is approved by one country, while another country or other countries provide scientific payloads, and then the data is shared. The Solar Orbiter, a collaboration between ESA and NASA, falls into the first scenario, where ESA develops spacecraft and NASA is responsible for the launch. China's Double Star Program (DSP), for instance, falls into the second scenario, and for the payloads on-board the two spacecraft of DSP, about 50% are from Europe. The advantages of this kind of cooperation include the cost reduction for an individual country as well as the sharing of scientific data and scientific results.

Finally, expanding the scientific output is also the motivation for international cooperation. Space science satellites missions will produce a huge amount of data when their in-orbit operations start. This is especially the case for Earth observation missions and solar observation missions because their observation targets are constantly changing and new data is generated every second. It is a tremendous task and almost mission impossible for a single team in an individual country to analyze all the data generated by these missions, so adopting the open data policy to obtain the maximum scientific output is the goal pursued by all institutions that manage this kind of space science mission. By opening up the entire process of the mission development and attracting international teams, it is possible to maximize the participation of scientists, which lays the foundation for future joint data analysis and expanding the science output. There are countless examples of this kind of collaboration. Almost all space science missions involve, from the beginning, international partners who bring new perspectives and ideas, as well as rich experience and lessons learned, to help mission teams better define mission requirements, improve payloads calibrations, program data processing software, and ultimately analyze data together and share scientific results.

13.3 Main Forms of International Cooperation

As mentioned above, it is imperative to establish cooperation in space science among different countries. However, since space science missions are invested by governments, the precondition for international cooperation depends more on the cooperative intention between the two governments than the cooperative intentions among the scientists [2]. Therefore, based on the closeness of the inter-governmental relation, international cooperation takes the following forms.

A joint workshop, like a loose form of cooperation, is only for the scientists. If the topic of the workshop is politically sensitive for bilateral relations, multilateral workshops could be organized to facilitate exchanges among the scientists and lay the groundwork for further cooperation. In the field of space science, China has established the mechanism of regular bilateral meetings with the USA, Russia, and other spacefaring countries or organizations, and organizes, through the International Space Science Institute in Beijing (ISSI-BJ), regular multilateral workshops and forums covering various disciplines of space science.

Another form of flexible international cooperation that does not involve hardware exchanges is joint data analysis. Chinese scientists have the necessary access to a large amount of open data produced by the space science satellite missions in the world. To use the data, you need to contact the data stakeholder (i.e., the team of the mission's Principal Investigator), and an explicit acknowledgment to the mission should be made upon publishing research results. Closer collaboration is a joint publication after intensive academic discussions. For the time being, the data of several Chinese space science satellite missions are already available to international research teams, such as Chang'e lunar exploration program, Zhangheng-1 (China Seismo-electromagnetic Satellite), and Hard X-ray Modulation Telescope (HXMT).

For closer partners, joint development of scientific instruments and providing reciprocal rideshare launch opportunities are common forms of cooperation. Such collaborations often do not involve any fund exchange, the data will be shared in a reciprocal way. When there is a fund involved, the collaboration usually becomes commercial in nature: the commercial revenue comes at the cost of losing the privileged equivalent access to data enjoyed by the science team, as defined by the data policy. Since this kind of collaboration involves hardware exchanges and rideshares, top-level support from the contributing governments in the form of a cooperative agreement is required. The cooperation framework allows the cooperation partners to obtain preferential terms in component tax exemptions from the customs.

A more intimate form of cooperation is joint projects at the mission level, such as joint development of satellites or providing launch opportunities. In general, the partners in this kind of cooperation already have in-depth cooperative relations, high-level mutual trust, and support. This is the closest partnership, but a signed cooperation agreement between the governments is still required because any delays or fund reduction from one country will affect the other. This kind of cooperation also does not involve any fund exchange, with the involved partners providing only hardware or rideshare launch opportunities.

A higher level of cooperation is inter-program or inter-mission cooperation. The IACG-coordinated collaboration, as mentioned above, on the joint detection of 1P/Halley upon its return to the inner solar system is an example. Such collaboration also requires close cooperation between governments but does not require any hardware exchanges. Intensive sharing and interaction happen primarily during the data analysis phase, so this kind of cooperation is relatively easy to achieve, despite the fact that it is higher-level cooperation.

13.4 Challenges

Of course, there are many difficulties and challenges associated with international cooperation. As mentioned above, the biggest challenge is mutual trust. Although the parties involved in the cooperation have the motivations, how to tackle the difficulties arising from the implementation of the cooperative projects remains an issue. In face of difficulty, to make collective efforts to overcome it or make unwarranted charges to each other is the touchstone of mutual trust. In order to avoid the scenario of ambiguous responsibilities, clear and detailed labor distribution and interface should be mapped out in the agreement signed at the beginning of the cooperation.

Another challenge in collaboration is that cultural differences can lead to mutual misunderstandings. When a problem arises, both parties need to put themselves in the shoes of the other culture to give a new understanding of the problem. For example, in some countries, the culture allows subtle expression, and there is a possibility that a person saying yes actually intends to say no because saying no directly is not deemed as a polite manner in that culture. In this regard, the characters and working styles of the managers and technicians directly involved in the collaboration are also important. It always happens that the members of an international team of a space science mission become lifelong friends after smooth and successful collaborations. The reason behind this is that they have overcome many difficulties together and have developed strong friendships in the process.

Despite the difficulties and tremendous challenges, the necessity for international cooperation in space science remains the main motivation to keep collaborations on track, which explains that a large number of ongoing space science missions have international cooperation to a certain degree. They simply differ in how close the cooperative relations are.

As for the agencies managing the space science missions, efforts should be made to actively promote and maintain international cooperation, establish a hierarchy of links at all levels among the countries, institutions, scientific teams, and technical teams, and facilitate cooperation in different forms such as joint workshops, data exchanges, piggybacked launches, joint missions, etc., so as to bring the wisdom of all mankind into full play and mobilize the most extensive resources in a bid to promote the development of space science.

References

1. Bonnet R, Manno V (2014) International cooperation in space: the example of the European space agency. Translated by Lei L. Science Press, Beijing
2. Zhou Y (2008) Perspectives on Sino-US cooperation in civil space programs. Space Policy 24(3):132–139

Chapter 14
Strategic Planning of Space Science in China

14.1 Introduction

Chinese Academy of Sciences (CAS) has been leading the space science strategic planning since 2006. Some strategic planning such as *Space Science & Technology in China: A Roadmap to 2050* [1], *the Medium and Long Term Plan of China's Space Science Missions (2020–2025)*, and *Calling Taikong—A Study Report on the Future Space Science Program in China* [2],have been released and published (as shown in Fig. 9.3). In addition, the National Natural Science Foundation of China organized the compiling work of *Space Science* [3] published in 2019, which introduces the development of space science disciplines. The *Calling Taikong*, as the comparatively complete space science plan of China in recent years, analyzes the science questions to be addressed with spacecraft and introduces in detail 23 science programs covering various space science disciplines. This chapter will focus on these programs.

14.2 Scientific Questions

Based on the workshops held in 2014–2015, China's space science community has reached a consensus on the following two science themes to be addressed. The first one is "How did the universe and life originate, and how do they evolve?", and the second is "What's the relationship between the solar system and human beings?"

© Science Press 2021
J. Wu, *Introduction to Space Science*, Springer Aerospace Technology,
https://doi.org/10.1007/978-981-16-5751-1_14

14.2.1 How Did the Universe and Life Originate, and How Do They Evolve?

This theme involves the following scientific questions:

How did the universe originate, and how does it evolve?

What is the universe made of, and how does it evolve?

What are the origins of the structures and objects of different scales in the universe, and how do they evolve?

Is there any new physics beyond the current basic physics theories?

How did life originate, and how does it evolve?

How did life originate, and how does it evolve?

Acquire evidence of life elsewhere.

What are the kinetic properties of matter, and what is the rule of life activity in the space environment?

What are the kinetic properties of matter in the space environment?

What is the rule of life activity in the space environment?

14.2.2 What's the Relationship Between the Solar System and Human Beings?

This theme involves the following scientific questions:

What is the nature of solar activities?

What is the nature of solar micro-phenomena?

What is the nature of solar macro-phenomena?

What is the origin and evolution of the solar system, and its relationship with the Sun?

How did the planets in the solar system originate, and how do they evolve?

How does solar activity transmit and evolve in interplanetary space?

How does solar activity affect the Earth's space environment?

How does the Earth system evolve?

How did the Earth system change?

Why did the Earth system change?

How will the Earth system change?

How will the Earth system science advance to adapt itself to global change?

The above-mentioned questions are deemed as the most fundamental and crucial questions by the science community. These questions could be a tool to inspire the science team to propose science missions. If one of the questions is answered, or even partially answered, it will contribute greatly to human's understanding of nature.

14.3 Mission Proposals

To answer the scientific questions mentioned above, space science programs are proposed as follows.

14.3.1 Black Hole Probe (BHP) Program

Through observations of compact objects such as black holes and gamma ray bursts, this program aims to study high-energy processes of cosmic objects and the black hole physics, and explore the extreme physical processes and laws in the universe with extreme objects, such as black holes, as probes of how stars and galaxies evolve.

14.3.2 Diagnostics of Astro-Oscillations (DAO) Program

For celestial bodies in the universe, the change of their electromagnetic radiation signals with time provides the basic information of their internal structures and activities. The periodic light variability of stars, white dwarfs, and neutron stars plays a vital role in our understanding of their nature. DAO program aims to conduct high-precision photometric and timing measurements of electromagnetic radiation at various wavebands and non-electromagnetic radiation (such as gravitational waves), to understand the internal structures of various astrophysical objects and the processes of various violent activities.

14.3.3 Portraits of Astrophysical Objects (PAO) Program

PAO program aims to directly acquire the portraits of stars, planets, white dwarfs, neutron stars, and black holes beyond the solar system, and high-resolution pictures of the central regions of galaxies, star formation regions, supernova remnants, as well as jets. It surveys the sky in various bands using deep imaging, and also provides high-resolution maps of the sky with background radiation at different wavebands. All of these will play a vital role in addressing scientific questions such as: what is the universe made of?

14.3.4   Spectroscopy of Astrophysical Objects (SAO) Program

SAO program measures the spectra of celestial objects with high resolution at various wave bands (main bands: visible, radio, and X-ray).

14.3.5   ExoPlanet Exploration (EPE) Program

EPE program aims to (a) search for and characterize Jupiter-like, Earth-twin exoplanets beyond the solar system; (b) precisely measure and systematically analyze the critical physical parameters of exoplanets including mass, orbit, and radius; (c) build the database about important physical parameters like planet radius, density, effective temperature, albedo, atmospheric environment, greenhouse gas, and surface gravity; and (d) address the question "whether there is another Earth in the universe".

14.3.6   Solar Microscope Program

By observing the Sun with high spatial resolution and multiple frequencies, Solar Microscope program aims to study the basic physical processes, such as the solar inner structure evolution, the origin of magnetic fields, the coronal structure and dynamics, and the trigger mechanisms of solar bursts and particle acceleration mechanisms.

14.3.7   Solar Panorama Program

Besides the high spatial resolution study for the Sun, Solar Panorama program pays more attention to the global behavior of the Sun, by multi-wave band diagnostics, to explore a physical connection between small-scale motion and the large-scale consequences.

14.3.8   Space Weather Chain Program

Space Weather Chain program focuses on the key coupling processes in the solar-terrestrial space. It studies the whole process and basic physics of formation, release, transfer, conversion, and consumption of the energy for large-scale disturbances in space weather events, in order to understand the mechanisms of how electromagnetic solar radiation and high-energy particles impact global climate change.

14.3.9 Micro-Sats Program

Micro-sats program aims to detect solar activities, interplanetary space, the Earth's magnetosphere, radiation belts and the ionosphere, the middle and upper atmospheres, and address key scientific questions such as the basic physical processes in space physics and the key areas, processes, and effects in space weather modeling and forecasting.

14.3.9.1 Look-into-the-Sky Program

Look-into-the-Sky program aims to improve the ability to monitor the geospace environment of the Sun, solar wind, magnetosphere, ionosphere, and middle and upper atmosphere. It arranges a ground-based monitoring network covering China's major aerospace bases and equipment test bases, major cities, and observation sites so as to build a middle or small-scale resolution monitoring capability, and unveil the regional characteristics of the space environment above China's territory as well as its relationship with overall global change.

14.3.9.2 Mars Exploration Mission

Mars Exploration Mission conducts detailed investigation about the global and local reconnaissance, in situ exploration, and sample analysis, through Mars global remote sensing, local reconnaissance, and fixed-point sample return. Its overall science goals include the following: (a) select the landing site and explore the conditions of life existence; (b) investigate Martian soil properties, water (ice), gas, and the composition; (c) detect Martian atmosphere and climate characteristics; (d) investigate the geologic characteristics, evolution, and comparative planetology. Through integrated detection to detailed analysis of local characteristics, and then to the internal composition research, it analyzes the surface processes, structural characteristics, geological units, and internal structure of Mars. On top of this, it studies the geological history of Mars and provides important inspiration for our studies of the Earth, especially the evolution of the Earth's environment, by comparing it with the Earth, the Moon, Venus, etc.

14.3.9.3 Asteroid Exploration Mission

Through flying alongside, touchdown, and sample return, Asteroid Exploration Mission aims to carry out near-Earth asteroid global reconnaissance and in situ exploration.

14.3.9.4   Jupiter System Exploration Mission

The overall science goals are to (a) study the structure of Jupiter's magnetosphere; (b) study Europa's atmospheric model; (c) investigate Europa's surface ice layer topography and thickness; (d) focus on solar wind structure among the Venus, Earth, and Jupiter; and (e) study terrestrial life in different space environments in terms of life condition, adaptability, and evolution.

14.3.9.5   Water Cycle Exploration Program

The water cycle is the dynamic process of water in a cycle driven by solar radiation, gravity, and other energies. It is the most active cycle process among the three major cycle systems (water, energy, and biochemistry) of the Earth. Water Cycle Exploration Program aims to investigate the links, mechanisms, and characteristics of water cycles connecting the Earth's hydrosphere, atmosphere, cryosphere, pedosphere, biosphere, and lithosphere, the internal matter and energy exchange, and the biochemical processes.

14.3.9.6   Energy Cycle Exploration Program

Energy Cycle Exploration program aims at investigating the impacts of the energy of solar radiation and the infrared energy budget of the Earth system on the movements and changes of the Earth system's main fundamental components (atmosphere, oceans, cryosphere, and ecology), where cloud in the atmosphere and its complicated interactions with radiation and aerosol play a significant role in regulating climate changes. The program carries out cloud-aerosol-radiation monitoring, earth radiation, and climate monitoring at Lagrangian points (L1/L2), lunar-based global change exploration, thermosphere probing, near-space atmospheric research, and atmospheric metal layer laser detection research.

14.3.9.7   Biochemical Cycle Exploration Program

Terrestrial and marine ecosystems play a crucial role in the global carbon cycle by means of photosynthesis, respiration, decomposition, as well as carbon release and absorption accompanying interferences such as fire disasters. Biochemical Cycle Exploration Program aims at the investigation of the impact of carbon release and absorption in terrestrial and marine ecosystems on the global carbon cycle. By observing the carbon cycle, soil moisture, and vegetation structure, the key functions of the ecosystem are to be studied.

14.3.9.8 Qingying Program

In Chinese, "ying" means "filled with fluid-like substances". Qingying Program carries out research on the fundamental science of microgravity fluid physics, applications and application fundamentals of microgravity fluid physics related to space exploration activities, and inter-discipline questions in microgravity fluid physics.

14.3.9.9 Qingyang Program

"Qingyang", cited from a well-known poem by chairman Mao Zedong, means "sailing into the sky" in the context of the poem. Qingyang program is mainly concerned with the processes in the micro-gravity environment, such as phase transition of materials, crystal growth, and material formation, as well as the physical and chemical properties of condensed matter system, and the processes and changes of thermophysical properties of melts.

14.3.9.10 Qingyan Program

In Chinese, "yan" means "flame". The three key research areas of Qingyan program are the mechanisms of turbulent combustion, coal combustion, and related heat transfer processes, and fire safety issues in the design and operation of spacecraft.

14.3.9.11 Space Fundamental Physics Program

Space Fundamental Physics program carries out verification of basic theories of fundamental physics with space tools.

14.3.9.12 Tengyun Program

"Tengyun", which literally means "cloud mounting" in Chinese, reminds readers of the heroic Monkey King "Wukong" from the classical epic *Journey to the West*. The novel is an exciting adventure where the hero battles against various monsters in order to get the scriptures. Tengyun program studies the phenomena, processes, and regularity of life activities under the special environment in space as well as the forms of terrestrial life in extraterrestrial space.

14.3.9.13 Taoyuan Program

In Chinese ancient literature, "Taoyuan", which literally means "the land of peach blossom", refers to the land of peace, away from the turmoil of the world. Taoyuan program aims to search for extraterrestrial and intelligent life and to study the origin of the universe, evolution, and basic rules of life. The program focus on future "Europa" or "Enceladus" missions, targeting those moons of Jupiter or Saturn which are likely to be covered by ice shells, underground oceans, and atmosphere. The possible life materials or forms in the atmosphere, ice shell, or seawater are planned to be sampled online and analyzed by orbiters and rovers.

14.3.9.14   Manned Space Engineering Science Program

This program includes space science research and applications in eight disciplines (31 subjects in total), such as space life science and biotechnology, microgravity fluid physics and combustion science, space materials science, space fundamental physics, space astronomy, space physics and space environment, and Earth science and applications. It may implement several hundred space science research projects. As for space life sciences and biotechnology, studies are carried out, on space fundamental biology, space life science frontier and inter-disciplines, space biotechnology and its application, space radiation biology, and bases on advanced space ecology and life support system. As for microgravity fluid physics and combustion science, the research covers microgravity fluid dynamics, two-phase flow, heat transfer with phase change and their application, and combustion science and its applied research. As for space materials science, the research includes the mechanism of synthesizing materials and thermal-physical property of materials in space, and on material manufacture in the case of major strategic demand. As for space fundamental physics, the studies are carried out on cold atom physics in space, Relativity and gravitational physics and the technologies involved, ultra-high-precision time-frequency system and its application, and low-temperature condensed matter physics experiment. As for space astronomy, the focus is on the fundamental science questions of "one black" (the black hole), "two darks" (dark matter and dark energy), and "three origins" (the origin and evolution of the universe, celestial bodies, and extraterrestrial life). It, mainly through sky survey, conducts high-precision multi-color photometry and spectral survey, dark matter particles and cosmic ray detection, variable objects and eruption detection, and the development of new space astronomy technologies. As for space Earth science and application, it covers the Earth science research related to global changes, new Earth observation remote sensor and its application, and environmental resources and natural disasters and related applications. As for space physics and space environment, it carries out research on space environment forecasting and monitoring and space physics detection.

14.4   Technologies

New technologies are needed to acquire new data. Some of the technologies are for special use, such as to improve the resolution of telescopes at some wave band, while some are for public use, such as small-size spacecraft, ultra-light large space deployable mechanism technology, and ultra-stable clocks.

14.4.1 Ultra-High-Resolution Imaging Technology

Ultra-high-resolution imaging is an important technology for space science high-precision observation. In order to improve the spatial resolution, in the future, on the one hand, a larger aperture of an optical telescope is needed, and on the other hand, interference technology can be used to realize synthetic aperture imaging, for example, large aperture interferometry imaging through formation flying.

14.4.2 Ultra-High-Precision Time Reference Technology

Ultra-high-precision time reference technology, the core technology of global navigation positioning system, originates from principles of space astronomy and has significant applications in global time synchronization, ocean monitoring system, low-orbit mobile communication constellation, satellite formation flying, etc. Meanwhile, it is the basic technology for space science missions such as solar system probes, VLBI, Earth gravity detection, verification of general relativity, and measurement of physical constants.

The technology mainly includes various new space atomic clocks, and long-distance high-precision time transfer and comparison technologies. Space atomic clocks include on-board atomic clocks and space ultra-high-precision atomic clocks. The former is applied in global navigation positioning systems and various satellites, with the advantages of small size, low power, and high precision; the latter, as the time reference in space, aims to calibrate various atomic clocks worldwide. Currently, the new generation on-board atomic clocks are Coherent Population Trapping Maser (CPT-Maser) and Pulsed Optically Pumped Maser (POP-Maser), with resolution up to 10^{-15}. The space ultra-high-precision atomic clocks include Ramsey microwave clocks, with a resolution up to 10^{-16}.

14.4.3 Distributed Satellite Constellation Technology

Distributed satellite constellation refers to a system composed of two or more satellites distributed on one or more orbits, cooperating with each other to accomplish a spaceflight mission. The satellites in this system are kinetically independent. The system increases antenna aperture and optic system aperture through interferometry measurement and sparse-aperture technology, to improve the overall performance of a space science mission, enabling exploration which is difficult for an individual satellite under current technology conditions. Distributed satellite constellation technology is also able to effectively lower the cost of a space science mission, enabling space exploration with much smaller size, lighter mass, and fewer resources. The technology includes control and measurement technology of position and attitude of formation flying, micro–nanosatellite system of in-orbit reconfiguration of sparse aperture imaging, multi-scale monitoring of solar-terrestrial environment based on separate payloads, space science detection system of four-satellite formation, and design technology of fly-around orbit for exploration of planets and their moons in the solar system.

References

1. Research group on space science & Technology of the Chinese Academy of Sciences (2009) Space science & technology in China: a roadmap to 2050. Science Press, Beijing
2. Ji WU (2016) Calling Taikong—a study report on the future space science program in China. Science Press, Beijing
3. National Natural Science Foundation of China, Chinese Academy of Sciences (2019) Space science. Science Press, Beijing

Chapter 15
Relations of Space Science, Space Technology, and Space Applications

15.1 Introduction

The previous 14 chapters cover space science history, science frontiers, space technologies, space engineering management, international cooperation, and strategic planning. This chapter will discuss the relationship between space science, space technology, and space applications, which will benefit our future work and avoid unnecessary disputes and misunderstandings.

15.2 Definition of Space Technology

Space endeavor begins with the technology development of rockets and satellites. Let's see what space technology is.

Space technology originates from the launcher technologies needed to enter space. By the early twentieth century, human's dream of escaping the Earth's gravity had been restricted on paper design or in fantasy. The first modern rocket was created by the U.S. engineer Robert Goddard. First, he created the first-generation rocket engine based on the combination of fuels (chemistry) and mechanical design. The pioneering members of the Caltech rocket group [1] in the 1930s were mechanical engineers and chemical engineers, while it was in the later phase that aerodynamicists, responsible for orbit computing, were involved. Then, with the development of electronics, control technology was introduced to better control the rockets. When the launcher was getting mature and satellite technology development was ongoing, the knowledge about the principles of electronics and physics was also inevitably necessary, such as energy, control, tracking telemetry control and communication, materials, semi-conductor components. When it came to manned spaceflight missions, space life sciences was introduced. Now space technology has evolved into an independent and comprehensive engineering technology category, involving chemistry, mechanics, electronics, physics, and space life sciences, as shown in Fig. 15.1. Mathe-

© Science Press 2021
J. Wu, *Introduction to Space Science*, Springer Aerospace Technology,
https://doi.org/10.1007/978-981-16-5751-1_15

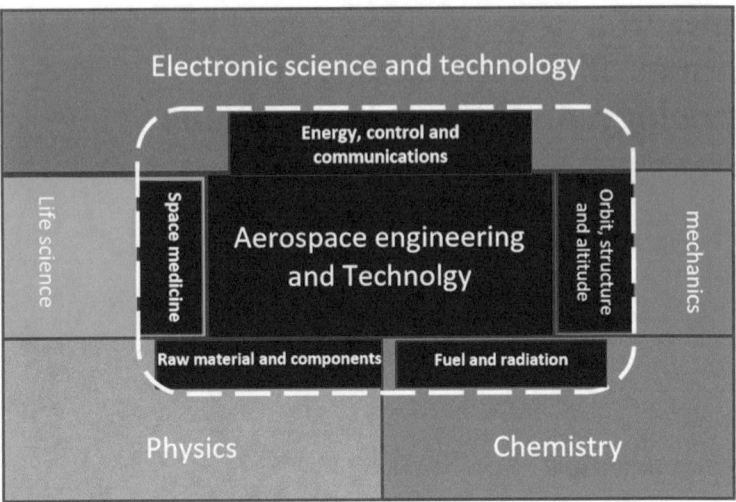

Fig. 15.1 Relations of space technology (space engineering technology) with related disciplines

matics is not mentioned specifically, since it is the foundation of all science disciplines and technologies.

Thereby space technology is defined as the comprehensive engineering technology aiming to explore, exploit, and utilize space and the celestial bodies excluding the planet Earth.

15.3   Definition of Space Science

The *direct* research on space originates from the study on atmosphere and gradually extends to outer space [2]. Hence the earliest space research dated back to the upper atmosphere science, and ionosphere research which started in the early twentieth century. After 1957, the radiation belt was found by satellite, which leads to the rapid development of space research, expanding from the middle and upper atmosphere physics, ionosphere physics, all the way to magnetosphere and interplanetary physics. Space physics (space plasma physics) became the earliest core discipline of space research or space science. Since the 1960s, satellites have been used to observe the Earth and the universe, and space technology began to intersect with Earth science and astronomy research. New disciplines of space science were thus created, such as space Earth science and space astronomy. Later, due to the development of manned spaceflight, microgravity science and space life sciences became the new input. Solar system exploration further significantly promoted the development of planetary science, with the involvement of astronomy, Earth science, and space physics. Space fundamental physics is the extension of fundamental research of physics. Currently, the disciplines of space science and their relations are shown in Fig. 15.2.

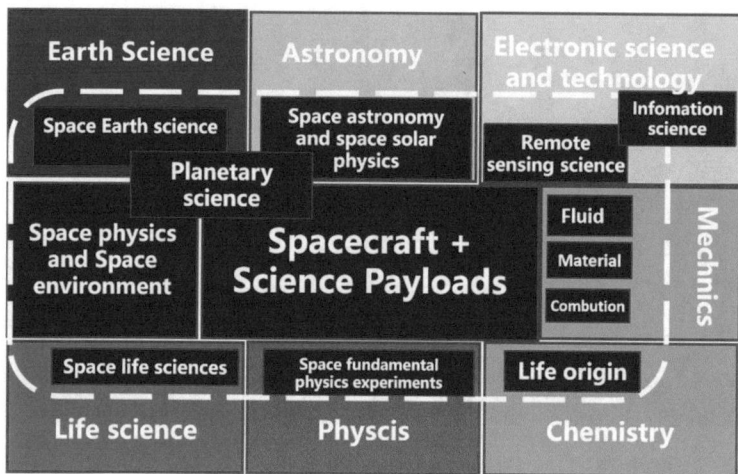

Fig. 15.2 Disciplines of space science and their relations

Thereby, with spacecraft as the main tool, space science is aimed to study natural phenomena and the underlying rules in physics, astronomy, chemistry, and life sciences which occur in solar-terrestrial space, interplanetary space, and even the universe as a whole.

15.4 Definition of Space Applications

Led by the development of space technology and space science, space applications emerged in the 1960s. The earliest applications showed up in the areas of satellite communication and Earth observation. Due to its broad coverage and the corresponding global communication capability, satellite communication was rapidly popularized, substituting shortwave wireless communication and wired telegram communication. Earth observation technology first found its applications in military reconnaissance, then extending to meteorology, ocean, city planning, land use, disaster monitoring, etc. Later on, with the further development of space science and space technology capability, satellite navigation gradually developed its applications, which is now the necessity for human activities. Based on the observation of the Sun and the geospace, human beings gradually realized that space weather events, triggered by solar eruption, should be blamed for the failure of spacecraft in orbit and some of the high-tech infrastructures on the ground, which makes space weather forecasting the important applications. Part of manned spaceflight experiments' achievements was transformed into development bases of new materials, seeds, and medicine, thus yielding new applications areas (Fig. 15.3).

Thereby space applications are a general term for the science and technology sector where their results are used to serve human society.

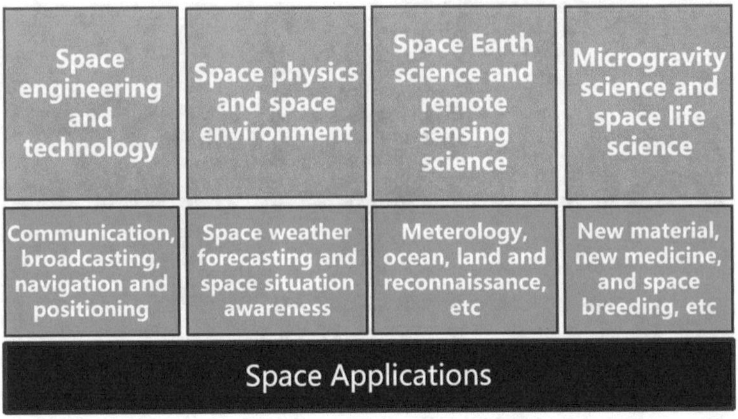

Fig. 15.3 Applications of space science and technology

15.5 Relations of Space Science, Space Technology, and Space Applications

After more than 60 years' development, space science, space technology, and space applications have become relatively independent sectors of national space activities. Meanwhile, they are very closely related to one another. First, due to its nature to explore the unknown, space science poses new demands for space technology, such as higher resolution, higher precision of altitude measurement and control, higher precision of orbit determination, and larger ratio of payloads to spacecraft. For engineers in space technology, the development of a space science mission is more challenging than that of a market-driven space mission. This is because when a market-driven space mission gradually gets mature, its performance parameters do not need changes, which makes the mission development into repetitive production with no more challenges for engineering designers. In addition, space science provides new theories for space applications. Once the new rules are discovered and new theories triggered by space science explorations develop related applications, they will be quickly adopted by the application department to serve the social and economic development.

Next, let's see space technology. The development of space technology depends mainly on mechanical engineering, control engineering, electronic information engineering, and systems engineering. Hence any breakthroughs in these areas, especially when led by applications with plenty of civilian users, such as chip technology and artificial intelligence, are able to promote rapid development of space technology. In recent years, microsatellite and CubeSat stand as good examples. The rapid development of space technology also inspires scientists to use new technologies in space science missions, to acquire new science data. Therefore, space technology supports and significantly promotes the development of space science and space applications.

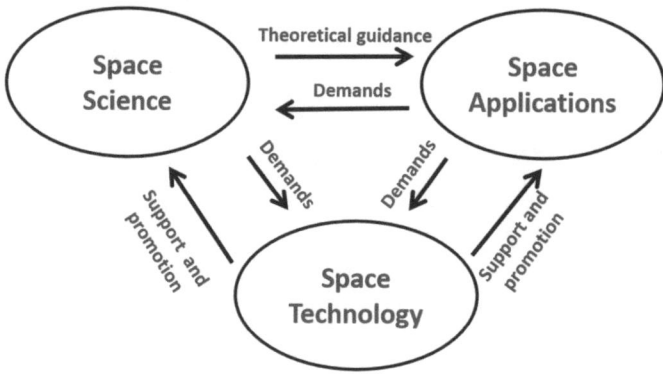

Fig. 15.4 Relations of space science, space technology, and space applications

Finally, we come to space applications. Space application, in fact, is the application of space science and space technology in social, economic, and military areas. Therefore, its development depends strongly on the development of space science and space technology. Further understanding of space and the Earth, and discoveries about the kinetic properties and rule of matter and life in the space environment, provides a theoretical base to guide new applications. New breakthroughs of space technology also serve human life in a time-efficient manner. In turn, space applications, with their development, pose new demands for space science and space technology. For example, the applications in Earth observation, to more accurately forecast weather and climate, pose constantly new challenges for space Earth science research. Another example is military applications, which focus on high spatial resolution and high reliability particularly, and its recent development requires accurate positioning and ultra-high detection ability, such as all-time and all-weather services, and detection of a covered object, even penetrating ocean surface to communicate underwater and detect underwater objects. These pose constantly new demands for space technology. Figure 15.4 summarizes the relations of space science, space technology, and space applications.

References

1. Harvey B (2019) China in space—the great leap forward. Springer Praxis Publishing, Chichester
2. National Natural Science Foundation of China, Chinese Academy of Sciences (2019) Space science. Science Press, Beijing

Printed in the United States
by Baker & Taylor Publisher Services